Méziane Boudellal
Power-to-Gas

T0074158

Also of interest

Chemical Energy Storage
2nd Edition
Schlögl (Ed.), 2022
ISBN 978-3-11-060843-4, e-ISBN (PDF) 978-3-11-060845-8,
e-ISBN (EPUB) 978-3-11-060859-5

Perovskite-Based Solar Cells.
From Fundamentals to Tandem Devices
Laalioui, 2022
ISBN 978-3-11-076060-6, e-ISBN (PDF) 978-3-11-076061-3,
e-ISBN (EPUB) 978-3-11-076065-1

Energy, Environment and New Materials Vol. 1–3
van de Voorde (Ed.), 2021
[Set ISBN 978-3-11-075497-1]

Vol. 1 Hydrogen Production and Energy Transition
ISBN 978-3-11-059622-9, e-ISBN (PDF) 978-3-11-059625-0,
e-ISBN (EPUB) 978-3-11-059405-8

Vol. 2 Hydrogen Storage for Sustainability
ISBN 978-3-11-059623-6, e-ISBN (PDF) 978-3-11-059628-1,
e-ISBN (EPUB) 978-3-11-059431-7

Vol. 3 Utilization of Hydrogen for Sustainable Energy and Fuels
ISBN 978-3-11-059624-3, e-ISBN (PDF) 978-3-11-059627-4,
e-ISBN (EPUB) 978-3-11-059410-2

Hydrogen Storage.
Based on Hydrogenation and Dehydrogenation Reactions of Small Molecules
Zell, Langer (Eds.), 2019
ISBN 978-3-11-053460-3, e-ISBN (PDF) 978-3-11-053642-3,
e-ISBN (EPUB) 978-3-11-053465-8

Méziane Boudellal

Power-to-Gas

Renewable Hydrogen Economy for the Energy Transition

2nd Edition

DE GRUYTER

Author
Dr. Méziane Boudellal
3 Rue Des Iris
31130 Balma
France
mbou@netcourrier.com

ISBN 978-3-11-078180-9
e-ISBN (PDF) 978-3-11-078189-2
e-ISBN (EPUB) 978-3-11-078200-4

Library of Congress Control Number: 2022950096

Bibliographic information published by the Deutsche Nationalbibliothek
The Deutsche Nationalbibliothek lists this publication in the Deutsche Nationalbibliografie;
detailed bibliographic data are available on the Internet at http://dnb.dnb.de.

© 2023 Walter de Gruyter GmbH, Berlin/Boston
Cover image: audioundwerbung/iStock/Getty Images Plus
Typesetting: Integra Software Services Pvt. Ltd.
Printing and binding: CPI books GmbH, Leck

www.degruyter.com

—

To my mother
To my family

Preface to the first edition

Transport, industry, housing and services sector: the "drivers" of these activities are the energies necessary to move, produce, heat or light, distract, for example. Electricity (which is not an energy source but an energy vector) plays an increasing role.

This trend, on the one hand, is due to the increase in requests of standards for buildings (both residential and commercial), leading to a reduction in heat requirements. On the other hand, development and extensive use of the Internet, computer and multimedia technologies, including office automation, have increased the number of equipment used and electricity consumption in some sectors.

Generation of electricity began in the nineteenth century and relied on hydropower, coal and natural gas. In the twentieth century, this generation was extended to other technologies such as nuclear technology. Consequently, there have been increased atmospheric pollution (CO_2, particulates etc.) and nuclear risks (not only at the plant level but also with regard to storage of radioactive waste).

The need to reduce CO_2 emissions has pushed towards creation of new "clean" electricity production paths, such as photovoltaic or wind power, with hydropower still in use. Although the number of facilities has exploded mostly because of subsidies, these facilities now allow for a significant production of renewable electricity, which can reach more than 50% of total electricity production, depending on the country and weather conditions.

Electricity production unfortunately has a major drawback: variability. This mainly depends on meteorological conditions such as wind and sunshine. Even if forecasts are fairly accurate, production maxima do not always coincide with consumption, resulting in higher than demand production; hence, there arises the need to be able to store this excess electricity for later use, regardless of its form.

Among the many technologies available, and depending on the country and the current or expected surplus volumes, which are sometimes large, the one that is going to emerge should allow the maximum of this surplus to be stored.

https://doi.org/10.1515/9783110781892-202

Preface to the second edition

Would this foreword had been written at the end of 2021, it would have confirmed the trends from the first edition. However, COVID and, most importantly, the war in Ukraine from the beginning of 2022 have mixed the cards. The disruption of the global economy due to the different lockdowns has led to tensions in the manufacturing industry, especially in Asia, resulting in factory shut down, reduced production and export, associated with the "recovery" after the COVID pandemic decreased in 2021.

The war in Ukraine with destructions, exodus and victims has also shown the too large energy dependence at different levels of many countries. It has also shown the lack of anticipation while relying for decades on cheap oil or natural gas, sometimes mainly from one country which is the case for Germany, for example.

Faced with the limitations of the stop of Russian natural gas deliveries and the stop of Russian oil import, the governmental responses of some countries have been to switch either to fossil fuels from other suppliers (oil or LNG from the USA, the Middle East and other countries) or to increase the extraction and use of coal for electricity generation (Germany, for example). Nuclear energy has also seen a comeback in the form of planned new plants (France) or keeping the ones that were supposed to shut down in activity (Germany), even if the shutdown was one of the Green Party's main criteria. In France, a coal power plant that was supposed to shut down in March 2022 was allowed to continue operation until (at least) March 2023. It will use 210,000 tonnes of coal imported from South Africa, North America and Columbia. The site was intended to be converted into "green" chemical production and large-scale hydrogen production. The urgency of the new start is also due to the critical situation of France's nuclear power plants: as of the beginning of October 2022, from the 56 available reactors, 27 were shut down due either to maintenance or corrosion issues.

The costs of planned investments (new nuclear power plants in France), subsidising gas or fuel prices, or even taking over bankrupt energy suppliers (like Uniper in Germany, nationalised in September 2022), will mean fewer investment possibilities for renewable energies. Relying on distant suppliers (through pipelines or vessel carriers) means also a less reliable source than local or close production. In September 2021, the pipelines Nord Stream 1 and 2 had leaks possibly caused by sabotage. The use of vessel carriers for oil, LNG or hydrogen can also be disrupted (bad weather conditions, for example) and is not a constant flow.

The economic aspects of the energy sector are strongly influenced by the high 2022 inflation, the latent COVID pandemic, the geostrategic situation in Ukraine and Taiwan and disruption in the logistic chain (especially from China) for the supply of goods and equipments. This results in very high energy, transportation and raw material prices.

Will the energy transition to a low carbon society be (temporarily) left aside or will we see in the mid-term a comeback and speed up to the initial objectives? The next decade will show the trends.

https://doi.org/10.1515/9783110781892-203

Contents

Introduction

Faced with environmental challenges (pollution, reduced fossil fuel reserves, nuclear waste etc.) and a growing demand for energy, there is the necessity of a sustainable solution. Renewable energies such as wind and solar are unfortunately subject to weather conditions and therefore to a variability of production. As a result, there already exists a gap between production and consumption today and it will increase in the future. It is therefore necessary to store unused electricity produced in order to be able to recover it during times of high demand or to use it in another form.

What technology should be used to store this excess electricity? The first solution that comes to mind is batteries. Although some have relatively large storage capacity, costs and limited lifetime do not allow for large-scale use. Other solutions exist, but all are limited in their storage capacity, compared to expected surpluses.

Apart from direct storage of electricity, the other approach is to convert it into another form to be able to store it. This is where electrolysis comes into play, i.e. conversion to hydrogen and oxygen, hence the name **power-to-gas.**

This hydrogen can be stored in different forms, converted into another gas (e.g. methane) or, if necessary, converted again into electricity in a fuel cell or a gas-fired power plant (or a cogeneration unit). Whatever the intended use, the excess electricity will be valued.

Although this approach seems to be the most elegant in terms of energy and technology, it still has some disadvantages. On the one hand, the cost of these installations is still high and, on the other hand, the hydrogen conversion capacity is still low compared to the current quantities of surplus electricity, and even more for those projected for the next decades.

Despite all, the power-to-gas technology combined with a decentralised approach (local use of hydrogen or methane) and optimised management at all levels, be it production, distribution or use (microgrid, smart grid and virtual power plant), would allow a better coverage of the needs as well as a facilitated stabilisation of the electricity networks.

The stakes of power-to-gas can only be seen in the context of the overall production and consumption of energy and their associated issues.

https://doi.org/10.1515/9783110781892-001

1 Global energy consumption

In recent decades, there has been a significant increase in global energy consumption in virtually all sectors, led by some countries like China and India, for example. Non-renewable sources of energy, which are still predominant, increase pollution, nuisances and greenhouse gas levels. Faced with the consequences on the population, fauna, flora and climate, it is necessary to turn to renewable energies.

One of the factors to be addressed before defining the potential of power-to-gas technology is the overall energy consumption, its evolution and the associated technical challenges (production, distribution and storage).

1.1 Strong growth in energy demand

The overall increase in prosperity leads to an increase in energy requirements, whether for transport, industry, tertiary or residential sector. The evolution of the gross national product is an indicator of the energy consumption, although a factor of uncertainty remains on the relation between these two parameters. Thus, projections for the next decades evaluate several scenarios.

1.1.1 Evolution of total energy consumption

Data for recent decades show a steady increase in consumption (Figure 1.1), even if it is weighted by climatic variations or economic crises. In 2020, the COVID pandemic and the associated lockdowns resulted in decrease of energy consumption of 4.5% for all sectors. However, 2021 saw a rebound with a growth of 5%, remaining slightly higher than 2019.

This energy comes from different sources: non-renewable sources such as oil, coal, gas and uranium and renewable sources such as wind, solar and hydropower. The contribution of each of these sources also varies from country to country (Figure 1.2).

A comparison of consumption trends shows that for many countries there is a tendency towards a decline or stabilisation of primary energy consumption (Figure 1.3) in recent decades. However, this does not mean that consumption by energy type or sector has also declined or remained stable for all countries, as shown in Chapter 2.

1.1.2 Energy storage

While solid or liquid fuels can be stored in large volumes, electricity storage is very limited in relation to production and consumption for technical (mainly relatively low capacity of available systems) and financial (high costs per kilowatt for some

https://doi.org/10.1515/9783110781892-002

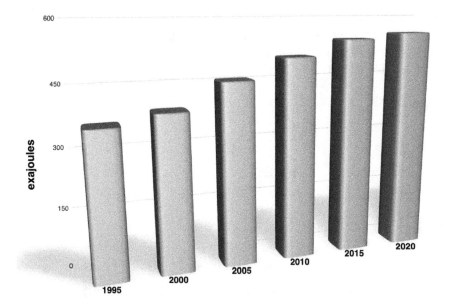

Figure 1.1: Global gross energy consumption in exajoules (BP Statistical Review of World Energy 2021).

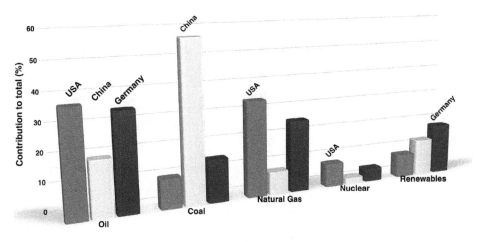

Figure 1.2: Contribution of different energy sources in Germany, China and the USA in 2021 (data: Our World in Data).

solutions such as batteries) reasons. The main electricity storage technology is Pumped Hydro Storage which represents over 90% of worldwide storage capacity (160 GW of power and 9,000 GWh of capacity in 2020). Electricity storage therefore remains very limited in relation to consumption (Table 1.1) or generation capacity.

The actual storage capacity of electricity is out of proportion in relation to production capacity or consumption when compared to that of natural gas or stored petroleum products. In the USA and China, the capacity of electricity storage accounts for about 1.5–2% of production capacity, with Japan being an exception, which is close to 6%. One could certainly include some of the hydropower plants as "latent" electricity, but they cannot be really considered as storage.

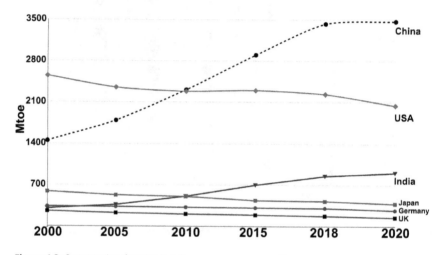

Figure 1.3: Comparative changes in primary energy consumption.

Table 1.1: Electricity storage capacities in 2021 (data: Countryeconomy.com, International Hydropower Association).

	USA	China	Japan	Germany	France	UK
Electricity production						
Yearly production (TWh)	4,048	7,601	889	545	523	299
Production capacity (GW)	1,143	2,200	348	248	136	113
Electricity storage						
Storage power (GW)	22	30.3	21.9	5.3	5	2.6

Security stocks for oil, petroleum products and gas

Each country stores gas and crude oil or petroleum products to meet either significant demand or a shortage. The IEA (International Energy Agency) is requesting and the European Union imposes, for example, a minimum stock of 90 days of net imports of petroleum products.

The EU has a natural gas storage capacity of about 113.7 billion m^3. Interconnections allow the exchange of this natural gas but there is no coordination between countries. The war in Ukraine in February 2022 took the EU by surprise as the filling level was unusually low. Even with a very high natural gas price (up to €345/MWh in March 2022 compared to an average of €25/MWh in 2021) countries had to buy natural gas to fill the storage for the winter 2022–2023 with an 80% filling target by November 2022 and almost 100% reached end of 2022.

Unlike the storage of electricity, gas, liquefied natural gas and oil or petroleum products stocks allow an autonomy of up to several months. The actual volumes in reserve vary according to management strategies, market prices (e.g. ideally purchases when prices are low) and level of consumption in relation to production or import. The maximum capacities of these reserves as well as the quantities stored are also constantly changing over the year, in view to ensure energy independence from potential risks (conflicts, very high prices) or respond to a high demand.

1.1.2.1 Electricity peak load

The management of peak loads (Figure 1.4) is critical for electricity suppliers. It depends mainly on the use of this electricity. Electrical heating is still widespread in some countries, and at very cold temperatures, demand is rapidly increasing. The same may happen in summer with the development of air conditioning. Electricity providers must be able to respond to them without delay. They have to balance real-time production and consumption to respond to these peaks.

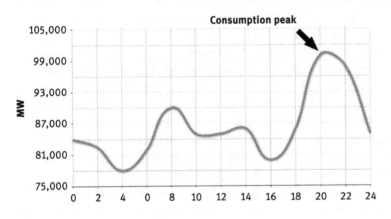

Figure 1.4: Example of peak consumption.

In case of demand greater than production, the low capacity of the electricity storage units requires either the use of imports if possible or run gas-fired power plants with a very rapid start-up time.

1.1.3 Energy consumption

Among the different countries, changes in primary energy consumption in recent years have experienced different trajectories. While in the long-standing industrialised countries (USA, Japan, UK, Germany etc.) the consumption has stabilised or decreased.

China for example does not yet show such a trend: it is still increasing (Figure 1.5). With a few exceptions, fossil fuels are still dominating.

Gross consumption and final consumption

The energy contained in raw sources such as oil, natural gas and uranium (primary energy) represents what could be used if the "extraction" yield was 100%. However, the different processes of transformation (coal, uranium or gas into electricity and/or heat) sometimes have significant losses (in a nuclear power plant only about one third of the initial energy is converted into electricity). The usable energy is called final energy.

1.1.4 Projection of the evolution of world energy consumption

In order to meet this growing demand, the strategies of each country are differentiated according to the available resources. Different models try to predict the evolution of consumption. Since the starting hypotheses are variable, at least two cases are often considered: the one with an evolution using the current mode of production and consumption more or less modulated (high hypothesis) and the other with an energy-saving or environmentally friendly approach (low hypothesis).

The "bp Energy Outlook 2022 edition" [1] presented three scenarios: *New Momentum*, *Accelerated* and *Net Zero* (Figure 1.6). *New Momentum* considers the evolution of the actual situation without drastic measures. *Accelerated* shows an approach of the energy system that should reduce considerably the CO_2 emissions leading to a global temperature increase of about 2 °C. *Net Zero* has a similar approach but is close to the 1.5 °C IPCC scenario with further possible improvements like carbon capture and storage. In all three cases, 2025 is the critical year: if no reduction is taking place by that date compared to 2020, the CO_2 emissions will not decrease significantly and global warming will not stabilise.

Another 2020 study [2] of the International Renewable Energy Agency (IRENA) is also in line with these forecasts at least in the trends to expect. It shows the necessity to increase the electrification of the energy system (electric vehicles, heat pumps etc.).

1.1.5 Potential for renewable energy sources

In a 2009 study, updated in 2022 [3], Richard Perez, one of the IEA/SHC experts (Solar Resource Knowledge Management), compared the different sources of energy to global consumption. It emerges that solar energy is, by far, the renewable source that not only could theoretically cover all human needs but also have no limit in time, at least as long as the sun is active. In this study, wind power can also make an important contribution to energy production (Figure 1.7).

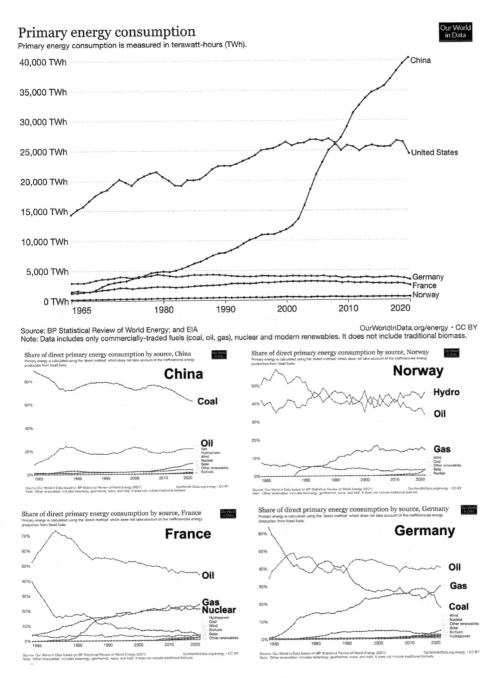

Figure 1.5: Evolution of energy consumption (Our World in Data).

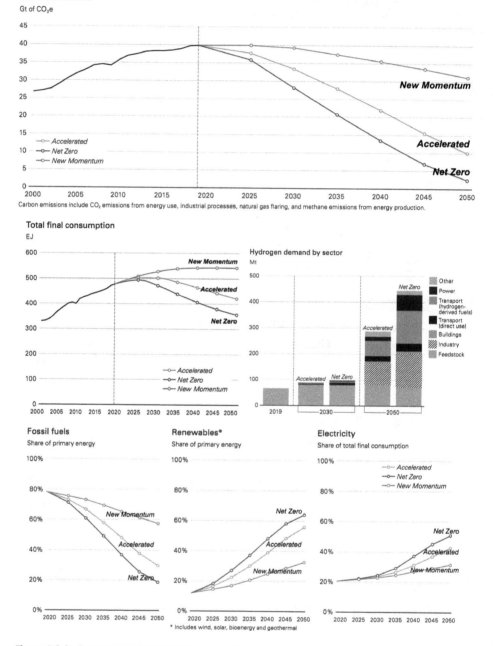

Carbon emissions
Gt of CO₂e

Carbon emissions include CO₂ emissions from energy use, industrial processes, natural gas flaring, and methane emissions from energy production.

Figure 1.6: bp Energy Outlook 2022 edition Scenarios.

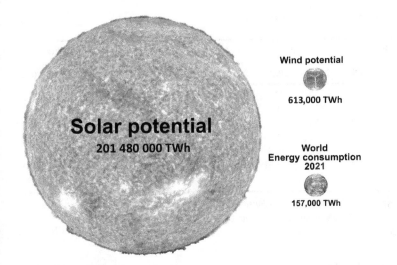

Wind potential

613,000 TWh

Solar potential

201 480 000 TWh

World
Energy consumption
2021

157,000 TWh

Figure 1.7: Potentially useful renewable energies (not to scale).

The annual solar energy reaching the surface of the earth is about 200 million TWh. Theoretically, it would be possible to produce 30,000,000 TWh of electricity annually from solar (photovoltaic or concentration with a 15% efficiency), 613,000 TWh from wind and 4,000 TWh from geothermal and hydroelectricity. World primary energy consumption in 2021 reached 13,500 Mtoe (about 157,000 TWh) or 1/1,300 of the total solar energy received!

1.2 Electricity production and consumption

In the residential sector, the increase in the overall standard of living and the desired comfort as well as the development of technologies, such as multimedia or the Internet, lead to growth or stabilisation at a high level of electricity consumption. Despite the reductions in consumption of various equipment, their multiplication partially (rebound effect) cancels efforts to reduce overall consumption.

1.2.1 Power generation

Electricity generation is still dependent on non-renewable energy sources, varying from country to country (Table 1.2).

These comparisons show not only the importance of coal but also the great dependence of some countries on a single technology such as France with nuclear (69% in 2021), for example, while the other countries have at least a second source of electricity

Table 1.2: Electricity production by primary energy (%).

	China		USA		Japan		Germany		UK	
	2016	2020	2016	2021	2015	2021	2016	2021	2015	2020
Coal	64	63	30	22	34	26.5	40	29	27	2
Natural gas	3	3	34	38	40	31.7	12	15.8	27	36
Nuclear	4	<1	20	18	1	5.9	13	12.3	23	16
Oil	5	<1	1	2	9	2.5	1	<1	1	3
Hydro	19	17	6	6.5	9	7.8	3	NA	1	2
Renewable	5	11	8	13.5	4	13.7	26	42.3	17	41
Others	NA	4	1	1	3	11	4	4	3.3	NA

Data: BDEW, Germany; United Kingdom Statistics Authority; IEA (values are rounded and totals do not necessarily correspond to 100% due to variations of data by source and rounding). For Germany, renewable data include hydro.

production (e.g. hydro for China and coal for Germany or the USA). Swedish production is mainly divided between hydroelectricity (45% in 2020) and nuclear power (29%).

1.2.1.1 New development or decline of nuclear energy?

By the end of 2021, 436 nuclear reactors were operational worldwide with a capacity of 389 GW. They produced about 10% of the world's electricity. Although many are under construction (52 units worldwide) or planned, especially in China (14 units end of 2021), by 2040 nearly 200 will have to be stopped (in 2016, 56% were over 30 years old and 15% were over 40 years old). The number of new reactors is still questionable due to delays in many construction sites (complex structures, increased safety measures and high costs), and at best, the percentage of electricity produced is expected to remain stable. The plants under construction benefit from artificially high subsidised prices: for the British plants at Hinkley Point C (the study of which was launched in 2007 and the construction decided in 2016), the government-guaranteed price of £92.50 is more than double the market price at that time. Operators are also faced with the high cost of updating old nuclear plants to bring them into line with increasingly drastic safety requirements (in France, they are estimated at more than €100 billion).

The war in Ukraine and the following energy issues (safe supply, storage etc.) has launched again in some countries a campaign to revive the nuclear industry either by launching new plants, postponing the decommissioning or extending the life of the existing ones. France for example has decided to build six EPR starting 2028 for an optimistic start in 2035. However, Germany that has planned to shut down the latest nuclear plants end of 2022 is still keeping this decision even if it has been delayed to 2023. Another technology is also booming: small modular reactors having less than 300 MW of power. Some countries are considering them, and start-ups (NuScale in the USA) are in the development phase.

The treatment of the hundreds of thousands of tonnes of nuclear waste created and to come will be even more crucial, even though no country has so far been able to dispose of them for a very long period of time. While the emphasis is placed by some operators and governments on nuclear power for the absence of direct CO_2 emissions, the resulting waste is not insignificant. In 2019 about 60,000 tonnes of nuclear waste were stored in Europe. Very few countries have a solution for the very long term of nuclear waste storage. Germany started searching for an adequate location in 1977. The *Gorleben* pilot site was abandoned in 2021 being considered as not appropriated. France started in 1990 the *Cigéo* project with a test site in Bure but no final location has been yet approved. Only Finland is building an underground storage facility to be operational by 2024–2025. It is supposed to be "safe" for 100,000 years after its sealing year 2120.

Electricity production also shows the relatively low efficiency of nuclear power plants compared with the corresponding primary energy, which generally does not exceed 30–35% (Table 1.3). Combined cycle gas turbine natural gas power plants have an efficiency reaching 60%.

Table 1.3: Nuclear primary energy and electricity consumption in France in 2021.

Overall primary energy	Nuclear primary energy	Final electricity production	Electricity efficiency
2,657 TWh	1,072 TWh	410 TWh (344 TWh from nuclear)	32%

Electricity generation is also only a part of the overall energy production, as shown in energy flow charts (Figure 1.8).

1.2.2 Increase of electricity consumption

Despite the gains in energy efficiency of consumer goods, electricity consumption is stagnant or increasing (Table 1.4), but at the global level it is still increasing.

A German study has shown that the factors responsible for this increase are population growth, but above all the increase in income (increasing expenditure on capital goods). This is confirmed by the ExxonMobil study (Figure 1.9) showing the increasing influence of the *improved standard of living* on electricity consumption.

1.2.2.1 Specific electricity

For some purposes, there is no alternative to electricity
- for residential sector: lighting, auxiliaries such as ventilation or pumps, appliances, computers and multimedia,

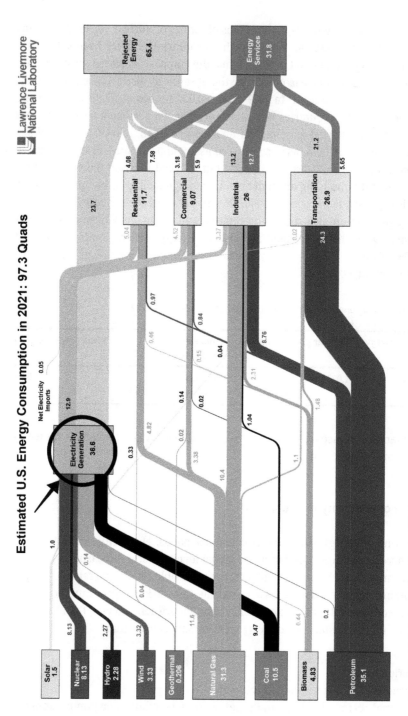

Figure 1.8: Global energy flows for the USA (Lawrence Livermore National Laboratory).

- for the tertiary sector: lighting, office automation etc. and
- for the industry: motors, machine tools, electronics, computers, etc.

Table 1.4: Increase in electricity consumption in TWh (data: World Energy Council, IEA, ENERDATA).

	1990	2000	2010	2020	Change 1990–2020
China	520	1,140	3,630	7,800	**15.0×**
USA	2,700	3,600	3,900	4,200	1.6×
India	270	380	730	1,600	**5.9×**
Japan	790	960	1,070	1,000	1.3×
Brazil	210	330	460	621	**3.0×**
Germany	455	500	550	575	1.2×
South Africa	165	208	257	235	1.4×
Australia	154	210	253	265	1.7×

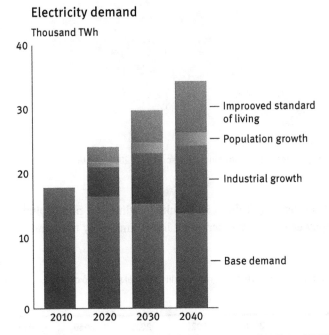

Figure 1.9: Factors affecting global electricity consumption (ExxonMobil).

The trend for the residential sector is caused by an increase in this specific electricity consumption due mainly to the multiplication of equipment, especially computing, multimedia and Internet (e.g. development of streaming).

Logically, electricity should be reserved exclusively for these specific uses. Other applications such as heating of hot water or other fluids can be covered efficiently by energies such as renewables, biogas, solar thermal or wood (pellets, for example).

1.2.3 Significant growth in electricity generation from renewable sources

In recent years there has been an increase in the production of renewable electricity with a steady increase of installed capacity (Figure 1.10). This same trend is to continue in the next decades.

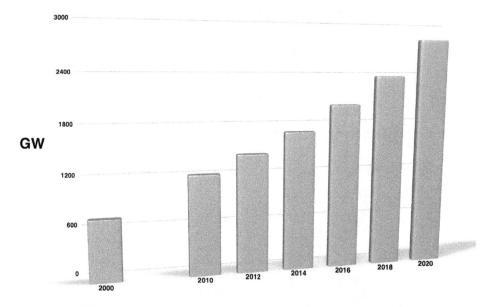

Figure 1.10: Global capacity of electricity generation from renewable sources in 2021 (IRENA, Renewable Energy Statistics 2021).

The global production is still increasing (Table 1.5), thanks to wind and photovoltaic electricity generation. Hydropower is also still increasing but is limited by the number of new appropriate sites.

Table 1.5: World growth forecasts of renewable electricity generation in TWh (data: IEA, ExxonMobil, IRENA, BP).

	2000	2010	2020	2025	2040	2050
Gross electricity generation	15,200	21,500	23,500	36,000	46,000	57,000
Renewable electricity	3,000	4,222	7,643	18,000	31,000	46,000
Included hydropower	2,650	3,443	4,418	6,000	7,000	8,000
Global energy production	116,300	145,400	160,500	172,000	167,000	164,000

1.3 Electricity market

Electricity networks are increasingly interconnected. The balance between electricity supply and demand is influenced by market exchanges where the price is fixed according to local or national criteria, such as surplus production or high demand. The exchanges are, depending on the legislation of the countries, either regulated by an organisation or left to the arbitrations between suppliers and buyers.

1.3.1 Electricity networks

All users require an uninterrupted supply of power and stable (voltage and frequency) variables. The different actors (production, Transmission and Distribution – T&D) must ensure that production and consumption coincide (Figure 1.11).

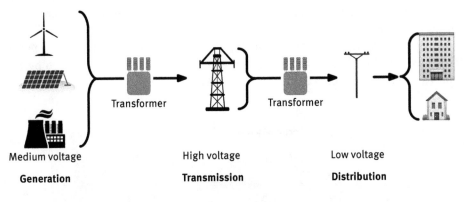

Medium voltage   **High voltage**   **Low voltage**

Generation   **Transmission**   **Distribution**

Figure 1.11: Power grid components.

Power lines are typically high/very high (50–750 kV), medium (6–30 kV) or low voltage (≤400 V). Electricity is predominantly transported and distributed as alternating current, but land-based or underwater lines carrying DC (high-voltage direct current – HVDC) are beginning to be built. Progress in the transmission of DC allows transportation over thousands of kilometres of high power at reduced costs and with low losses (transportation of 6 GW under a voltage of 800 kV over 2,000 km with less than 3% losses).

To measure the quality of a network, the indicators used are SAIDI (System Average Interruption Duration Index), which indicates the average annual outage duration or SAIFI (System Average Interruption Frequency Index). However, SAIDI does not reliably take into account all interruptions, those that are less than 3 min, nor the microbreaks.

1.3.2 Network stability

The criterion for the management of electrical networks, apart from the production/ consumption ratio, is their stability. This is measured by the frequency variation with respect to the theoretical basic value (50 or 60 Hz).

Electricity producers (nuclear, gas, hydropower or wind farms, photovoltaic) and carriers must be able to regulate this frequency. If production and consumption balance, the frequency is stable. The inertia of the production system affects the frequency variation: if consumption is greater than production, the frequency decreases and vice versa.

If one of these two variables increases or decreases, the frequency will change. If the frequency is lower, it is necessary to increase the power supplied (production or import) or decrease the demand and, if higher, to reduce power to the grid (decrease production or increase export) or increase the load. This frequency is stabilised, but it can vary within a defined range.

In Europe, the Union for Coordination of Transmission of Electricity is responsible for defining the control of this frequency range, which must be maintained by the various operators.

Any variation must be corrected quickly. For this purpose, producers or suppliers define levels of control (primary, secondary and tertiary) that are represented by the required reserves, according to the time of use (Figure 1.12).

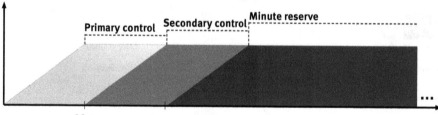

Figure 1.12: Correction of frequency changes-reserve activation.

The stability of networks involves different power stations and strategies (Figure 1.13):
– the base load stations provide on a continuous basis the minimum level of demand, generally estimated for 24 h (depending on the country, today it is mainly the nuclear or coal-fired plants that have this role)
– medium-load plants correct the difference between the base load and peak loads and operate only part of the time (gas-fired plants or eventually wind farms or photovoltaic)
– the peak load units operate only for short periods of time, but have a very short start-up time and a wide range of power variation (gas or hydropower plants)
– for local solutions, batteries of high capacity (several MW) are also used

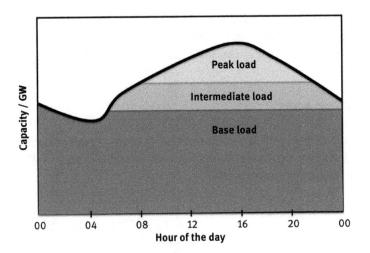

Figure 1.13: Load curve for a typical day.

1.3.2.1 Gas-fired power plants and security of supply

These power plants, which allow network regulation, especially for peak demand, have been undermined since the explosion of shale gas production in the USA. It has led to the fall in the price of coal and a massive export to Europe for use in coal-fired power plants, also favoured by the fall in CO_2 prices. Many operators put their gas-fired power station under cocoon or even plan to close it, which would entail a risk to the stabilisation of the power grids as those plants can be started very quickly.

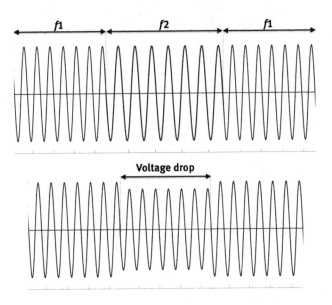

Figure 1.14: Example of variation of mains frequency or voltage.

The voltage supplied to the end user has a pure sinusoidal form of constant amplitude. Disturbances such as sudden change in load can vary the frequency (frequency excursion) or the voltage (drop) and possibly destabilise the network (Figure 1.14).

1.3.3 European electricity networks

The various European national networks are interconnected in order to facilitate trade (export of surplus or import). This network is the largest in the world (Figure 1.15). The exchange capacities vary from country to country.

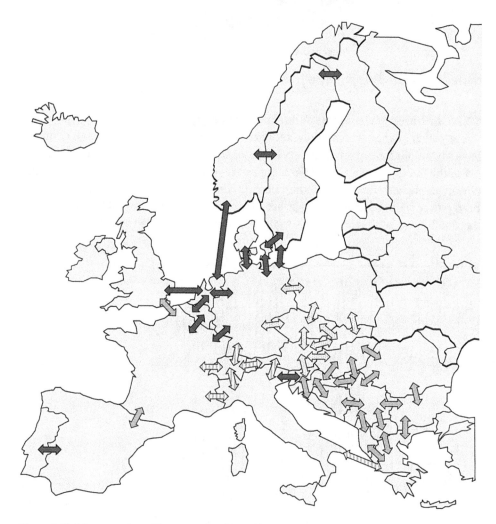

Figure 1.15: Interconnection of European electricity grids (ENTSO-E).

Those interconnections (400 in 2022) must ensure security of supply and meet national deficits when demand exceeds production. These exchanges also allow regulation of the electricity market due to competition. Europe, through the ENTSO-E (European Network of Transmission System Operators for Electricity) regroups 39 members from 35 countries in 2022.

1.3.4 North American electricity network

The USA has extensive trade with Canada and Mexico. The US and Canadian networks are completely interconnected. The USA is divided into four zones: Eastern Interconnection, Western Interconnection, Texas Interconnection and Alaska Interconnection (Figure 1.16). The Canadian provinces are grouped into three zones: Western Grid, Eastern Grid and Quebec Grid, including Atlantic Canada.

Figure 1.16: The US electricity grid (EIA).

Proposed in 2009, the project called Tres Amigas should connect the three major networks, i.e., 48 states with 8 Canadian provinces. The hub located in the state of New Mexico, point of convergence of the three networks, should allow the transfer of a power up to 20 GW. The node equipped with direct current (HVDC overlay) connecting AC networks should stabilise the network and make it more resistant to blackout. As of 2022, the project seems to be progressing but has been scaled down in terms of costs and technical infrastructure. However, different studies have shown the advantage of a single US unified electricity network.

1.3.5 Other networks

In Central America, the SIEPAC initiative includes Panama, Costa Rica, Honduras, Nicaragua, El Salvador and Guatemala. It resulted in the creation of a regional electricity market (MER) and the construction of a transmission infrastructure to increase the exchange capacity. Guatemala network is connected to Mexico.

The ASEAN (Association of Southeast Asian Nations) Centre of Energy (ACE) is since 1999 an organisation representing the interests of the 10 ASEAN countries in the energy sector. The second phase (2016–2025) of the ASEAN Plan of Action for Energy Cooperation concerns the development of sustainable energy.

1.3.6 Chinese World Global Network

Presented by the Chinese government in 2015, the Global Energy Interconnexion (GEI) envisions a worldwide electricity network. Supported by the GEI Development and Cooperation Organisation it targets in 2050 a global network of 177,000 km covering all continents and using UHT DC lines (800–1,100 kV). The first step was supposed to interconnect through 12 UHT lines North Asian countries including Russia and North Korea. However, the political situation resulting from the war in Ukraine has put the project on hold.

The European Commission has also been involved and proposed three different connecting routes: the north one crossing Russia, the south one crossing Afghanistan and the central route avoiding the two countries. Main issues are the evolution of the political situation and the costs: the GEI project was estimated in 2018 to €11,000 billion.

1.3.7 Need for a stable network

The increase in electricity generation from renewable and intermittent sources (wind and photovoltaic) is a challenge for electricity producers and distributors. For network stability, which is solved in many countries where electricity production can sometimes reach or exceed 100% of consumption, the possibility of storing surplus electricity in large quantities can be a stabilising factor. The need for large-volume storage technologies, i.e., the **power-to-gas** technology can play a role in this security of supply.

1.4 Electricity market structure

The electricity market is divided into production, transportation and distribution. Depending on the country, these activities are carried out by one or many operators. These three sectors are not necessarily covered by all operators and depend on the old network structures and policies of liberalisation and deregulation.

In Germany, for example, very large producers (E.ON, EnBW, Vattenfall and RWE) as well as dozens of municipalities and small operators are active. In the USA, the market is covered at around 50% by regulated, vertically integrated structures (production, transmission and distribution – T&D), while the rest is divided among a large number of producers or distributors. A very small fraction belongs to the government or directly to the consumers. For the other extreme, in France a single operator covers more than 75% of production and controls indirectly T&D.

In Europe, the primary distribution network (HT) is usually provided by regional operators. Germany and the UK, for example, are divided into four zones (Figure 1.17). Geographical breakdown ensures competition that optimises transportation costs, which is not the case in all countries (in France T&D is still a state monopoly). Since the 1990s, the European Commission has set up an internal energy market with the creation of ENTSO-E. The European electricity market has been fully open to competition since 1 July 2007. Countries such as France, Belgium, Germany and Holland (grouped as Central West Europe) have also formed a coupling of electricity prices leading to price convergence.

In the USA, the Federal Energy Regulatory Commission was created in the mid-1990s to introduce more competition into the electricity market. However, limited government regulation is necessary to control the conditions of competitiveness and possibly intervene. This gave rise to independent operators (Independent System Operators) that control, coordinate and monitor the electricity grid in a specific zone (in unregulated wholesale market areas). According to the Energy Information Administration, more than 60% of the electricity goes through these two groups. Public utilities, cooperatives or major cities cover about 15% of the market.

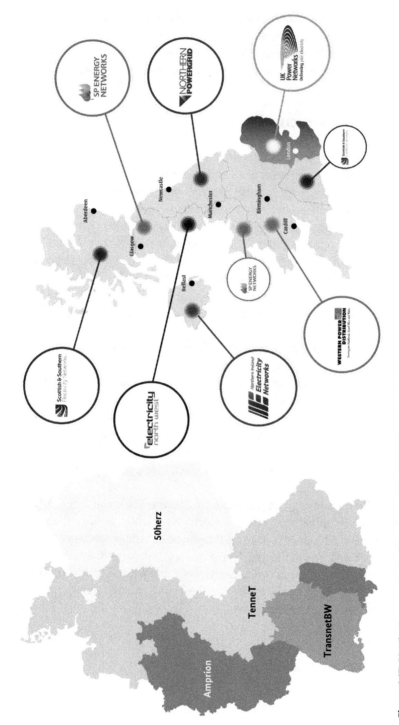

Figure 1.17: Distribution operators in Germany and the UK (BMWi, National Grid).

1.5 Structure of electricity prices

The wholesale electricity market is mainly divided into two options:
-   over-the-counter agreements that represent the majority of transactions
-   electricity exchanges

At European level, prices are negotiated according to the time criteria (Figure 1.18), de-pending on the period of the transaction. Prices can be fixed in advance (several days, weeks or months). The spot price (short term) is either D-1 (day ahead) or the day of the transaction (intraday) and varies according to supply and demand.

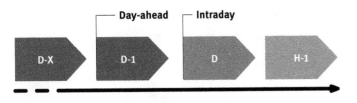

Figure 1.18: Electricity price adjustments.

These prices are traded on various stock exchanges, usually in several countries (the EPEX SPOT – European Power Exchange – for example manages the electricity markets for France, Germany, Switzerland and Austria). In the very short term (H-1), there are still possible adjustments.

Transactions can also be traded days, months or even years ahead (Figure 1.19).

If supply is higher than demand (a situation that occurs especially when photovol-taic or wind generation is important), then prices may even be negative: the supplier pays the buyer. According to ENTSOE-E data, the occurrences are gradually increasing from 925 in 2019 to 952 in 2021 (2020 was an exception with 1923 occurrences due to the low demand during lockdowns).

In Germany there were 139 h in 2021 with negative prices (day-ahead market) with an average price of €−16.4/MWh.

The example in Figure 1.20 is that of the German market in 2020–2021 with a large renewable electricity production resulting in negative electricity prices.

1.5.1 High electricity prices for consumers

Until the end of 2021, wholesale prices (Figure 1.21) were stable or declining while for consumers the price of electricity was increasing. In 2021, the average residential price in Europe was 23.69 cents/kWh. In some countries an important increase took part (UK and Greece) whereas some other countries decreased the price (Hungary, Netherlands, Norway) either by reducing taxes or due to falling costs.

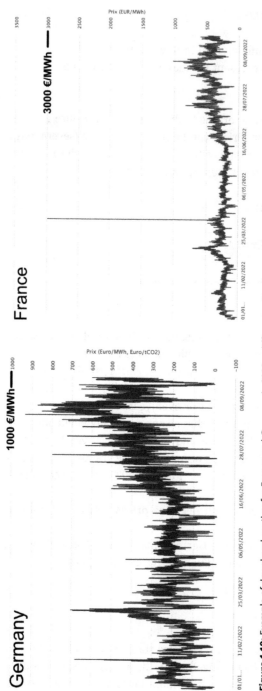

Figure 1.19: Example of day ahead auction for France and Germany in 2022 – different scales (energy-charts).

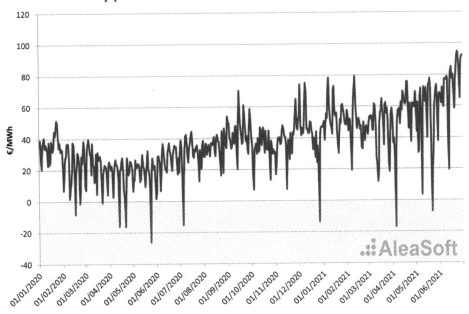

Figure 1.20: Negative electricity prices on the EPEX market for Germany (AleaSoft).

But following the Russian invasion of Ukraine in February 2022 electricity and gas market prices were peaking at never seen levels. Consumer's prices are also increasing consequently leading some countries to reconsider pricing policies (subsidies or limited increase).

1.5.1.1 High subsidies for non-renewable energies

Industrial countries continue to subsidise fossil fuels directly or indirectly. According to the International Institute for Sustainable Development [4], the G20 countries spent an average of US $584 billion annually through direct or indirect subsidies. The year 2021 saw a drastic increase: the OECD study covering 51 countries (OECD, G20 and 33 other countries) estimated the subsidies to producers and customers to US $697.2 billion (362.4 in 2020).

These data only concern oil, coal or natural gas. The nuclear industry also benefits from substantial financial support. The Euratom as part of the Horizon Europe Program will spend €1.38 billion for the period 2021–2025. In 2022, the European Union voted the "taxonomy" that includes certain nuclear and gas investments considered as "environmentally sustainable" economic activities. Two weights, two measures? Faced with electricity from renewable sources, subsidised directly by the user, these government subsidies can only favour non-renewable energies and polluting technologies.

Germany Electricity Spot Prices

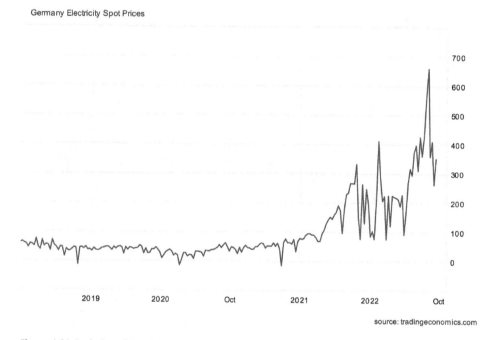

source: tradingeconomics.com

Figure 1.21: Evolution of the electricity spot price for Germany (trading economics).

1.6 Renewable electricity: a necessity

The electricity production from fossil fuels (coal, oil, natural gas or uranium) uses finite resources, which are also important for other sectors (chemicals, for example). Electricity can be produced from renewable sources. The need to increase this production is not only economic but also ecological (making non-renewable raw materials last, no CO_2 emissions or other pollutants or waste to be stored or recycled as for nuclear).

However, the war in Ukraine requires a new modelling of the energy sector in all aspect (production/generation, transport, exchange and use) including electricity.

References

[1] bp Energy Outlook, 2022 edition.
[2] IRENA, Global Renewables Outlook, 2020 edition.
[3] Perez, R. and Perez, M. Update 2022 – A fundamental look at supply side energy reserves for the planet, Solar Energy Advances, p. 1–7, Volume 2, 2022, 100014.
[4] International Institute for Sustainable Development. Doubling Back and Doubling Down: G20 Scorecard on Fossil Fuel Funding.

2 Electricity of renewable origin

Wind and photovoltaic have undergone significant development since the 1980s, due in part to the fact that these energies are (almost) free as far as "fuel" costs nothing and is available in abundance. Various government subsidies have also played an important role in this development. Today, the weak point of electricity from renewable sources (except mainly hydro and geothermal) is the production not related to consumption, depending on meteorological conditions (wind and sun). Hence today's need for optimised management, combined with available non-renewable sources (coal, natural gas and uranium), require to balance production to meet the demand.

2.1 Technologies

Renewable electricity generation is an alternative to non-renewable sources for many reasons:
- "fuel" does not have to be imported like oil or natural gas,
- enables increased energy self-sufficiency,
- reduces balance of payments (less trade deficit) and
- promotes industrialisation (wind turbine or tower production plants, electronic control).

Specificity of electricity of hydraulic origin: the technologies that will be presented are mainly characterised by their intermittence and variability, which is not the case for hydropower or geothermal energy, for example. The hydropower stations can be controlled (started upon request) and their power is controlled by varying the flow of water. If they can contribute to the stability of the electricity grid, they also have another role like watering for the agriculture, maintaining river water level and they cannot be emptied like a battery.

In terms of capacity, in many countries, unlike wind or photovoltaic, hydropower cannot be significantly expanded, as the potential of large resources has often been exhausted.

2.1.1 Wind energy

The wind energy production is divided into land (onshore) and marine (offshore) installations (Figure 2.1), often grouped together in parks of several tens or hundreds of wind turbines.

"Nuisance" of facilities: wind turbine installations often encounter protests from residents or environmental associations (nature protection, fishing etc.). Criticisms of

https://doi.org/10.1515/9783110781892-003

Figure 2.1: Offshore wind farm (Siemens).

wind turbines include non-integration in the landscape, birds' hazards and noise. Studies have shown, however, that these fears are often exaggerated and that localisation outside migration areas significantly reduces risk. Noise of onshore farms is covered by the wind noise.

It is also necessary to compare the few hundreds of wind turbines to the hundreds of thousands of electric pylons and associated power lines that disfigure the landscapes and also produce electromagnetic fields, sometimes installed even in inhabited areas.

2.1.1.1 Wind turbines

The power of the wind turbine generators is increasing. In 2022, Siemens Gamesa commercialised a 11 MW model with a rotor diameter of 200 m. Siemens Gamesa and GE Wind Energy had prototypes with 14 MW power (220 m rotor diameter), Vestas presented a 15 MW model with a 236 m blade diameter for a total height of 280 m, and the Chinese company MingYang Smart Energy installed a 16 MW turbine with 236 m rotor diameter (80,000 MWh/year). Each offshore 14 MW wind turbine can produce up to 74 GWh per year, representing the consumption of 16,000 households (with an average of 4500 kWh/household).

Repowering

This process involves replacing old turbines with new more powerful power. This amounts for the same number to increase the power supplied or to reduce the area occupied for an equivalent power.

2.1.1.2 Wind power generation

According to the Global Wind Energy Council 2022 report [1], the global installed capacity of wind power in 2021 was 849 GW (12.4% more that 2020 – Table 2.1). It should be noted, however, that this is a maximum theoretical capacity. World average production in 2016 represented about 28% of theoretical production and 33–43% for offshore wind turbines in Europe.

Table 2.1: Global capacity installed at the end of 2021 (in GW).

	Europe	Asia	America	Africa	Others	Total
Germany	64					
Spain	28					
UK	27					
France	19					
Others Europe	87					
China		344				
India		40				
Others Asia		30				
USA			134			
Canada			14			
Other America			41			
Total	**225**	**414**	**189**	**7**	**5**	**840**

2.1.1.3 Wind farms

At the beginning of 2022, more than 250,000 wind turbines were installed worldwide, including 120,000 in China, 70,000 in the USA and 28,000 in Germany. The USA, China and Germany dominate in terms of onshore facilities (Table 2.2).

Table 2.2: Main onshore wind farms in 2021.

Name	Power (MW)	Country
Gansu	8,000	China
Alta Wind Energy Centre	1,548	USA
Muppandal Wind Farm	1,500	India
Jaisalmer Wind Park	1,064	India
Fosen Wind	1,057	Norway
Shepherds Flat Wind Farm	*845*	USA
Meadow Lake	801	USA
Roscoe Wind Farm	781	USA
Önusberget	753	Sweden
Horse Hollow Wind Energy Centre	735	USA
Capricorn Ridge Wind Farm	662	USA
Fântânele-Cogealac Wind Farm	600	Romania

Offshore wind is growing more and more in spite of technical difficulties. Europe, with the majority of worldwide farms representing a generation capacity of 24 GW in 2021 (35 GW in the world), has still the largest number with a total of 5,800 wind turbines predominantly in the North Sea, the UK being the leader with 12.7 GW (Table 2.3). Many parks with a power higher than 1 GW are under construction like Dogger Bank A, B and C (the UK) with 1.2 GW each, Hornsea Two (the UK) with 1.3 GW or Ijmuiden Ver (the Netherlands) with 4 × 1 GW.

Table 2.3: Main European offshore wind farms in 2021.

Name	Power (MW)	Country
Hornsey 1	1,218	UK
Borssele 1&2	752	Netherlands
Borssele 3&4	731	Netherlands
East Anglia One	714	UK
London Array	630	UK
Gemini	600	Netherlands
Gwynt y Môr	576	UK
Greater Gabbard	504	UK
Anholt	400	Denmark
BARD Offshore 1	400	Germany
Global Tech I	400	Germany

In other countries, offshore wind parks number is increasing. China has an installed capacity of 26 GW in 2021 vs. 1.6 GW in 2016 and the USA inaugurated the first park (Block Island Wind Park) only at the end of 2016 but has an installed capacity of 42 GW in 2021.

2.1.1.4 Projections

In the different forecasts for 2050, installed capacity should increase drastically in order to meet the climate goals (CO_2 emissions and global warming) set by the different governments or organisations (Figure 2.2).

If wind power has benefited from large subsidies such as a guaranteed price for electricity produced or a premium for kWh, this approach has changed. In early 2017, German energy supplier EnBW won a bid for a 900 MW farm without any subsidy or kWh premium. This has been attributed to technical developments (high-power wind turbines with higher efficiency) and the best financing conditions. This technology being now proven, this trend has been confirmed since.

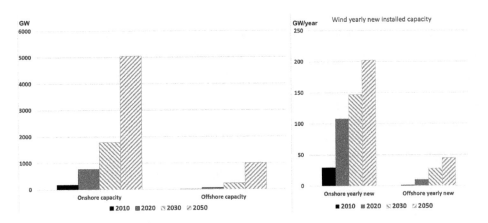

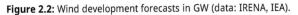

Figure 2.2: Wind development forecasts in GW (data: IRENA, IEA).

2.1.2 Photovoltaic

Photovoltaic solar energy can be divided into three segments: large-scale parks (Figure 2.3), installations on the roofs of industrial, agricultural or tertiary buildings and small installations on residential roofs. Photovoltaic is thus the only technology that can be easily implemented on a small scale (that of a residence).

Figure 2.3: Photovoltaic park (Siemens).

The contribution and development of each of these sectors, especially for the residential sector, varies from country to country and year to year (buying price of kWh variable and decreasing). In the US the installed capacity was distributed between residential (19%), commercial (10%), industrial (2%) and power utilities (68%).

The world generation capacity installed in April 2022 was 1,000 GW: 268 GW in China, 107 GW in the US and 165 GW in Europe (25.9 GW more than 2020: 56 GW in Germany, 22 GW in Italy, 17.9 GW in Spain and 14.6 GW in the UK).

Solar power shows a steep increase in installed capacity and power. In Europe, the installed capacity could increase in the most optimistic scenario from 165 GW in 2021 up to 1,050 GW in 2030 (Data from "Solarpower Europe").

The largest solar parks are located mainly in China and India (Table 2.4).

Table 2.4: Large photovoltaic parks in 2021.

Name	Power (MW)	Country
Golmund Desert Solar Park	2,800	China
Bhadla Solar Park	2,500	India
Longyangxia	2,400	China
Benban Solar Park	1,300	Egypt
NP Kunta Solar Park	1,200	India
Sheikh Mohammed Bin Rashid Al Maktoum Solar Park	1,030	Dubai
Jinchuan	1,030	China
. . .		
Topaz Solar Farm (California)	580	USA
Gemini (Nevada) *completed in 2023*	690	USA
. . .		
Witznitz *completed 2023/2024*	600	Germany
Nunez de Balboa	500	Spain

Some parks (Kozani in Greece with a power of 204 MW) use bifacial panels that collect light on both faces. The yield is higher than one-sided panels (up to 27% more energy).

To reduce the occupancy on the ground, floating parks are installed on lakes or reservoirs of dams such as Dezhou Dingzuhang in China with 320 MW, CECEP also in China with 70 MW and Sembcorp in Singapore with 60 MW. Asia has 85% of installed world capacity in 2022. In India, some irrigation canals are covered with photovoltaic panels that also reduce evaporation. The state of Gujarat in India has several achievements to its credit: the most important has a length of 3.6 km for a power of 10 MW and a project of 100 MW is planned (canals from the Narmada River). Other locations such as tunnel roofs can also be used: in Belgium, a rail tunnel is covered with panels with a capacity of 4 MW.

Photovoltaic in space: Space-based solar power consists of orbiting photovoltaic power stations transmitting electricity produced by laser or microwave to land stations. The first projects dating back to the 1960s, were mainly evaluated in the USA and Japan. As of 2022, no project has been started. But in September 2022, Airbus presented the *Power Beaming* project with no detailed technical specifications. The New Zealand Emrod company developing wireless power transmission (dating back to Nikola tesla, end of the nineteenth century), showed a small ground demonstrator using a 1.9 m diameter antenna for the transmitter (5.8 GHz) and the receiver separated by 36 m. It powered a model city, an electrolyser and a beer fridge.

Issues for space installation are the global efficiency (from PV modules to usable power), the size of the ground "receiver", the accuracy of the beam positioning to avoid any interaction with neighbouring areas etc.

2.1.2.1 Projections

The drop in the price of photovoltaic modules (US $0.19/W in 2020 vs 0.37/W in 2016) has led to the development of large photovoltaic farms and has an influence on the prices of electricity produced. However, since 2020, the price has started to increase (US $0.27/W in 2021/2022) due to COVID supply chain disruption, high demand and higher energy prices after the invasion of the Ukraine beginning of 2022.

According to the International Energy Agency (IEA) or the European Photovoltaic Industry Association, installed capacity is expected to grow extremely rapidly in the coming decades. The IRENA estimates that an installed capacity of 2,840 GW in 2030 and 8,519 GW in 2050 would be needed to meet the target of a decarbonised energy.

If the drop in the price of panels has expanded the market, the production of photovoltaic electricity remains limited. Its contribution to global electricity generation is about 3%. However, the electricity mix varies from country to country: 15% in Honduras, 8–9% in Spain and Germany and 2.8% in the US in 2021.

2.1.3 Solar concentration

In the Concentrating Solar Power (CSP) technology, mirrors reflect solar radiation either to the top of a tower or to a receiver where this energy heats a fluid (water, oil etc.) that drives an alternator (Figure 2.4).

One of the advantages of this technology is that it can operate in the absence of sunlight (also at night) if during the day part of the heat produced is stored in molten salts, for example, and recovered at night to operate the turbine. If it was used as early as the nineteenth century, the first work on important industrial installations dates from the years 1960 to 1970. At the end of 2021, installed capacity was 6.6 GW (5.02 GW in 2016). Spain with 2.36 GW dominates this sector, followed by the USA with 1.8 GW. The Spanish installed capacity has not significantly increased since 2013.

In China, end of 2021, there were 9 operational installations with a power between 50 and 100 kW for a total of 500 MW (original target was 1.3 GW for 2020) and 11 are planned for 2024.

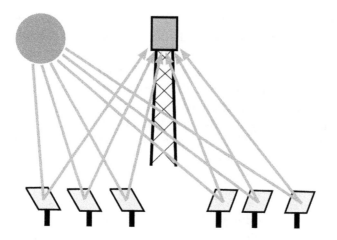

Figure 2.4: Diagram of a concentring solar installation.

However, all projects are either not successful or realised. The US Crescent Dunes CSP, located in Nevada, was operational in 2013. It had a power of 110 MW and 10 h of heat storage in molten salts. It was expected to produce 500 GW/year. In 2019 it was shut down due to low performance. In Chile, the Cerro Dominador CSP with a power of 110 MW was planned in 2014 but went operational in 2021. Its 10,600 mirrors and the 252 m tower allow a molten salt storage of 17.5 h and a 24 h/7 d operation.

The Gemasolar plant near Seville, Spain (Figure 2.5) uses 2,650 mobile mirrors to concentrate the beams towards a tower of 140 m. A salt (nitrate) is heated to 565°C and produces steam operating a turbine with an electrical power of 19.9 MW. Thanks to the stored molten salt, the autonomy without sun reaches 15 h and the plant can operate continuously several weeks 24 h a day with an average annual production of 110 GWh.

Inaugurated in 2016 the first phase of the Noor power station in Morocco had a generation capacity of 10 MW. The third phase brings the total power to 580 MW.

This technology is mainly used in areas of strong sunshine. Another important limitation is the high cost of such facilities faced to the drop in photovoltaic module prices. The forecasts for 2030 estimate, depending on the scenarios, an installed capacity of about 14 GW.

2.1.3.1 Solar chimneys

This technology, explored since the beginning of the twentieth century, also uses solar thermal energy but produces electricity by circulating air, heated in a greenhouse, from the base of the tower to the top, thus activating a turbine connected to a generator. This concept of the German engineer Jörg Schlaich was experimented between 1982 and 1989 with a prototype of 50 kW in Manzanares, Spain, with a tower of 195 m

Figure 2.5: Solar power station Gemasolar (Andalusian Energy Agency).

and a greenhouse of 46,000 m^2. Although numerous studies and simulations were published and projects announced, no significant project has been implemented.

2.1.3.2 DESERTEC

This concept, developed in 2003, was meant to provide a renewable electricity supply (wind, photovoltaic, CSP) located in northern Africa and the Middle East and transmission lines (Super Grid) to Europe. The total planned production was more than 600,000 TWh/year, which could cover the needs of these countries and of Europe. This project was extended to southern Europe and then to other desert areas in the world. However, given the important investments and the political implications, no practical progress has been made and the project stopped. Another initiative Desertec Industrial Initiative (Dii) started in 2013. The project has been "updated" with emphasis on hydrogen production and export.

2.1.4 Marine energies

A variety of different projects involving several technologies are being evaluated around the world. The marine energy used can be that of waves, tides or currents. Another form is the thermal energy or the osmotic effect.

Marine current turbines are submarine "wind" turbines located where the sea currents are high or in the course of a river. The existing equipment is either attached to a tower anchored to the sea floor or to a frame placed on the sea floor. In Japan, IHI Corporation has installed in 2017 at a depth of 50 m a 100 kW floating turbine (Figure 2.6) at Kuchinoshima Island (Kagoshima prefecture).

Figure 2.6: 100 kW floating current turbine (IHI Corporation).

The various constraints associated with this approach mean that potential marine areas (currents >2 m/s and geographical accessibility) are still limited. The experimental units (France, UK and Norway) are still of relatively low power, generally not exceeding 2 MW with a high cost per kilowatt. The production is very predictable, given the regularity of tides or currents. Numerous companies are either improving the turbines (Orbital Marine Power/2 MW), entering into the competition (Scottish Simec Atlantis Energy/500 kW, Verdant Power/US) or developing new concepts (tidal kite of 100 KW from Minesto/Sweden, hinged raft from Mocean Energy/UK).

The dream of the ocean currents: The Gulf Stream in the Atlantic or the Kuroshio off Japan and other ocean currents represent enormous energy potential. However, the distance to the coast and the depths to be reached are prohibitive obstacles. However, many projects, even being utopic, exist to use this gigantic potential.

Tidal energy is operational in only four countries: South Korea (Sihwa Lake, 254 MW, 2011), France (Rance, 240 MW, 1966), Canada (20 MW) and China (5 MW). It requires an appropriate site, heavy investments and may lead to risks to the ecosystem.

The exploitation of **wave energy** has attracted interest in dozens of projects. However, the experiments to date have not yielded convincing results to move to large-scale commercialisation. One of the first tested systems (Pelamis Wave Power, 750 kW) had the structure of an articulated snake. Others use floating buoys anchored at the bottom of the sea and connected to a generator (The Australian company Carnegie Wave Energy with its CETO 6 model 20 m in diameter capable of producing 1.5 MW; Figure 2.7) or oscillating panels. These systems encounter technical difficulties related to sea conditions (corrosion, storms) or technology (mechanical problems). Companies are still working on improvement of the reliability and yields (CETO increased the generated energy by 27% in 2021 by a better control). The program EuropeWave PCP is supporting seven different companies (CETO in this sector for demonstrators in 2022).

Figure 2.7: CETO 6 floating buoy (Carnegie Wave Energy).

The **ocean thermal energy** (OTEC) conversion can be exploited using the temperature difference between surface water (26–30 °C in inter-tropical zones) and deep water (4 °C to 1,000 m). These two sources at different temperatures can operate a thermal machine using an intermediate fluid (e.g. Rankine cycle). The yield is low (3–4%) but this technology can be a solution for islands with continuous production. The first power plant was built in Cuba in 1930, a 1 MW unit (OTEC-1) installed on a tanker was operational in Hawaii in 1980 and showed the feasibility of the concept. In Martinique, a floating unit of 10 MW planned for 2018 has been stopped due to high costs. In 2019 about 30 units were operational. The majority were low power (<20 kW). The most powerful was in Hainan, China with a power of 2 MW.

Osmotic energy uses the osmosis principle, which consists in bringing saltwater on one side of a membrane and freshwater into contact on the other. The water crosses the membrane and creates an overpressure that can operate a turbine. A 4 kW unit was built in 2009 in Norway by Statkraft but work was stopped in 2013. The Dutch company REDStack has been operating a unit of 50 kW since 2014. It is planning to upscale it to 1 MW. The French company Sweetch Energy is planning for 2023 a unit at the mouth of the Rhône River.

Limited potential of marine energy: All the solutions proposed are facing two main issues. The high costs, both for investment and maintenance, and also for the kWh produced (the US Department of Energy estimated the kWh cost at US $2.80 in 2019) as well as technical difficulties due to the maritime environment. While many countries have evaluation programmes underway and the overall theoretical potential for electricity generation is very large, the global contribution to electricity generation capacity is negligible: in 2021, 39.6 MW of tidal energy and 24.7 MW of wave energy were operational worldwide.

2.1.5 Biomass, geothermal energy

While being an indirect source of electricity production, the renewable origin of raw materials (biomass) or terrestrial energy (geothermal) makes it possible to classify electricity production as renewable. However, the possibility of managing the electricity generated (power, time of operation and running time) does not make it a variable and intermittent source such as solar or wind.

2.1.6 Renewable electricity production records

The increase in renewable power investments allowed to increase the share of electricity generation. In 2021, some countries reached the 50% mark: 46% for Germany (Figure 2.8), 58% for Portugal, 47% for Spain, 53% for Denmark and 98% for Costa-Rica.

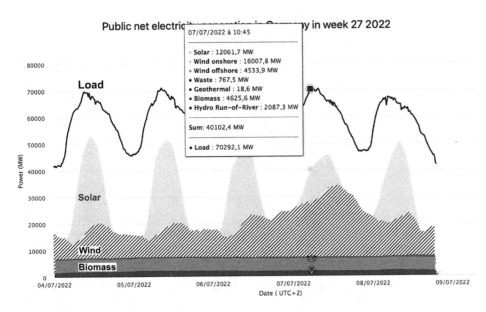

Figure 2.8: Electricity production in Germany week of July 2022 (Energy-charts.de).

2.1.7 Comparison of electricity costs by origin

The price of electricity produced according to the primary energy used varies depending on the development of technologies and their extension. If the price of solar or wind electricity was initially higher than the average price from non-renewable sources, then it decreased as new facilities were developed.

Grid parity: It is the price of electricity generated by renewable sources compared to that of non-renewable sources. For some countries (Germany and Southern Italy), this parity has already been or is reached. For solar photovoltaic systems, for example, this means a lower kWh cost than the network, hence the advantage of using this electricity directly without injecting it into the grid.

Levelised cost of energy: This criterion for comparing the costs of different energies considers all those involved, and among others, variables such as initial capital, operating costs, fuel and maintenance. These costs are estimated for the lifetime of the facility. Many recent studies have compared the costs of different sources of electricity generation (Table 2.5). They present a cost range for each technology showing similar trends, but the overall trend is a drop of the price of renewable electricity.

The costs of photovoltaic or wind power will continue to decrease as the number of installations increases (lower solar modules and wind turbine costs), while one of the other sources, especially nuclear power, will tend to increase (ageing of installations, maintenance costs, depollution etc.).

Table 2.5: Electricity price comparison (2021 data).

Study	Gas	Coal	Nuclear	Wind onshore	Wind offshore	Solar PV
Lazard-US (US $/MWh)	141–204	61–157	128–207	26–50	66–100	28–221
IEA (2019 data)	88 (CCGT)	130	150	56	75	56
Fraunhofer (€/MWh)	114–289	104–200	NA	39–83	72–121	31–110

2.1.8 Energy transition and renewable energy

As part of an energy transition, many countries have set targets for reducing the use of non-renewable sources for energy production. According to the timetables, the deadlines are 2025, 2030 and 2050. Not all countries, however, put the needed financial or legislative means to achieve these objectives. For many countries, however, this energy transition is possible. Numerous studies show the feasibility of generating up to 100% renewable electricity from 2030 to 2050.

In July 2021, the EU package "Fit for 55" has been adopted with the objective to reduce by 2030 by at least 55% the emitted greenhouse gases (GHG) compared to 1990 levels. The proposed series of measures concern among others vehicle CO_2 emissions reduction, decarbonisation of buildings through an improved energy efficiency, at least 40% of renewable in the energy mix, greener fuels for aviation and ships. This all associated to a social climate fund.

In April 2021, President Biden announced an US target to reduce GHG emissions by 2030 by at least 50% compared to 2005 values. This goal will be reached by investments in infrastructure and innovation to reduce pollution from the transportation and industry sector. A carbon-free electricity should be reached by 2035 for example.

In October 2020, Japanese Prime Minister Yoshihide declared that the Japan would reach net zero GHG emissions. However renewable will not cover 100% of generation: hydrogen, ammonia, nuclear and "clean" thermal power plants will be involved. Electricity use will be promoted in all sectors.

Many other countries have set objectives to either reduce GHG emissions or to reach net zero emissions by 2050. However, there are many challenges awaiting those countries: need of disruptive technologies at a large scale, very slow decrease of number of internal combustion engines. The main limiting factor is the fact that those objectives are mostly set by and for developed countries, including China. But the reduction of global GHG concerns ALL countries and many do not have either the technologies or the resources to move rapidly in this direction.

German concept "Energiewende"

Many countries are planning an energy transition but often without a clear global concept and/or a series of coherent long-term measures, often also with little binding and sometimes unreliable targets. "Energie-wende" is the German concept started in 1998 with concrete objectives. In 2011, after the explosion at the Fukushima reactors, Chancellor Angela Merkel decided supplementary measures like the phase-out of nuclear power by 2022. This has resulted in many reflections and accomplishments to accompany this waiver and chart new paths, such as improving energy efficiency and power-to-gas technology. This upheaval has allowed the German industry and research to be at the forefront in many energy sectors. However, until beginning 2022 no effective governmental measures to ensure conformity with the setup objectives (reduction of CO_2 emissions and use of fossil fuels, renewable energy boost. . .) has been really adopted or implemented: the German weekly *Der Spiegel* from 13 August 2022 had a cover page showing a wreath with the title: "here rest our climate objectives". Thus when, after the invasion of the Ukraine, supply of Russian gas declined and energy prices soared, a large consensus to speed up the energy transition was then met. The real involvement of the German government and industries is still open.

The lack of anticipation to look at other alternatives when depending mainly on a single supplier is one of the characteristics but not only for Germany. As an example, for the German industry is the ammonia/urea supplier SKW Stickstoffwerke in Wittenberg, Germany: in the week-end issue 35 of the German newspaper *Welt am Sonntag* (28 August 2022), the SKW Stickstoffwerke which are cracking the Russian natural gas to produce hydrogen said that they will see an increase of the natural gas cost of about €30 million! Such an investment would have allowed a Power-to-Gas installation able to produce (eventually) green hydrogen and be less dependent on imported natural gas.

The rise of renewable electricity: All studies and projections show that photovoltaic will become the first source of renewable electricity in 2050, followed by wind. Other sources, despite technological developments, are expected to remain limited due to high cost, complexity or (relatively) reduced power and will remain limited in their use as a response to specific needs. On the other hand, the renewable electricity would allow to eliminate the need to use non-renewable sources (oil, natural gas, uranium) since the initial energy (wind, sun) is free of charge. Moreover, the price of equipment (wind turbines, photovoltaic modules) should continue to fall. Maintenance costs are not to be compared with conventional power plants, not to mention nuclear ones with radioactive waste management and ageing.

2.2 Variability in production and electricity consumption

A significant volume generation of renewable electricity does not necessarily occur when demand is high, but rather depends on weather conditions (Figure 2.9). Total production/consumption management is therefore more critical because either the system must absorb quantities of electricity that can bring it to its maximum transmission capacity, or it is necessary to stop or reduce the power of certain power plants or even worse disconnect some wind or photovoltaic farms.

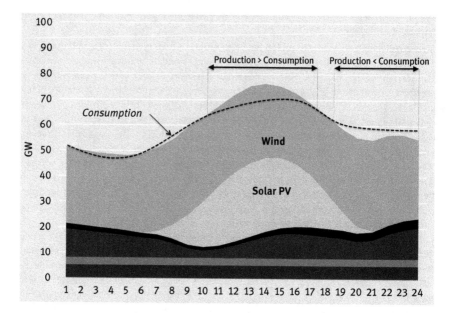

Figure 2.9: Example of a shift in renewable production/consumption (data: Agora Energiewende).

2.2.1 Production and consumption forecasts

Weather forecasts (sun, winds) are an important factor in the management of renewable electricity flows in the overall production framework. These forecasts, especially those of strong winds or storms, are also critical for the safety of wind power installations, for example.

2.2.2 Electricity flow management

The importance of production/consumption management will increase with the growth of electricity generation from renewable origin.

All power grids are concerned because the production areas, especially for offshore wind, are often far from the consumption areas needing more transmission lines on long distances. Ideal management also contributes to the reduction of electricity cost through optimisation of storage (less required) or of the load of transmission lines. The storage capacity must reduce the need for local, regional and international interconnections to export or import electricity by reducing losses.

2.2.3 The need for storage

The storage of electricity makes it possible to transfer the availability of this energy over time (Figure 2.10): storage in case of low demand or high production and recovery in the opposite cases. Even if storage capacities are limited, they increase the flexibility of the energy system and reduce the building of new transmission lines, especially if the storage is local.

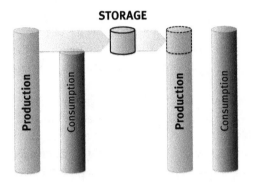

Figure 2.10: Advantages of storage for network regulation.

Direct electricity storage technologies are still expensive or limited in capacity and one of the objectives is to reduce these costs or develop an alternative solution(s).

Storage is related to the capacity (*Capacity Value*) of the generation of renewable electricity to correlate production and consumption. Photovoltaic solar energy has a significant capacity value when peaks in production and peaks in consumption coincide (e.g. in Spain). For Denmark, it will be wind power that will have a significant capacity value.

Renewable power: a challenge for production and transport: The intermittent generation of electricity from renewable sources increases the stresses on the electricity grid, which must be able to manage peaks of production that are not correlated with demand, even over short periods of time. Only storage allows optimisation of the energy efficiency of the various sources of electricity production. However, many countries (Spain, Portugal and Germany) have shown that high levels of electricity from renewable sources, sometimes as high as 100% (in Portugal or Denmark for example), can be integrated without disturbing the network.

2.2.4 Estimation of surplus electricity

The amount of renewable electricity produced will increase. Their variability will lead to periods when production will be greater than consumption. One solution would be to export this electricity to neighbouring countries if they are not faced with surpluses and if the capacity of the exchange lines is sufficient. For large surpluses, the only solution not to lose them remains storage in any form whatsoever.

2.2.5 Simulations

This is a prospective forecasting exercise, which yields results with a wide range of uncertainty, depending on the sources.

Many studies have tried or try to estimate the detailed annual surplus of renewable electricity. The difficulty is for a given period, to predict the maximum. These data are important because they determine the maximum storage capacity (e.g. hydrogen conversion) to be planned.

An increasing amount of surplus renewable electricity: The significant development of wind and photovoltaic power expected in the coming decades will lead, when the weather conditions are favourable (sun, wind), to volumes of electricity that the market will not be able to absorb due to a lower demand. These quantities may represent up to the equivalent of several days or even weeks of consumption and should be recovered for direct or indirect use.

2.3 Electricity storage

The need to make production and consumption of electricity to coincide with the increasing volume of renewable electricity requires an important storage in the event of overproduction. It can also stabilise or regulate the grid. Depending on the priority placed on storage, storage capacity or available power will be emphasised.

2.3.1 Why store electricity?

Electricity is produced and used in a continuous process where it is necessary to balance supply and demand. With renewable electricity generation increasing, networks could be sometimes at the limit of the power to be transmitted, shortages or blackouts due to technical problems or inclement weather can occur. The balance can be achieved only with a minimum of storage.

Storage allows:
- grid frequency regulation (50 or 60 Hz) with rapid response and for a short time (usually up to a few minutes)
- transition between production units requiring longer start time
- electricity generation management to store the surplus

2.3.2 Characteristics of a storage system

What are the important criteria for choosing a storage solution? Among those to be examined are:

- storage capacity
- reaction time
- charging and discharging time
- frequency of call
- number of charging–discharging cycles
- performance
- lifetime
- price of stored kWh

2.3.3 Storage technologies

There are many available technologies for storing electricity. They are technically differentiated by the principle involved (Figure 2.11), which affect storage capacity, efficiency, volume, lifetime, conversion method to electricity, level of maturity and cost.

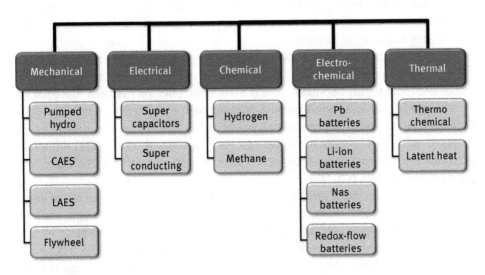

Figure 2.11: Electrical storage technologies.

2.3.3.1 Pumped hydrostorage

This technology stores electricity indirectly. A pumped hydrostorage station (PHS) consists of two large-capacity water reservoirs, at different altitudes, of pumps and turbines (Figure 2.12).

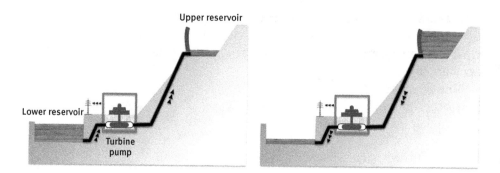

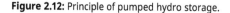

Figure 2.12: Principle of pumped hydro storage.

When the price of electricity is low, it is used to run the pumps bringing the water to the upper reservoir. When the demand exceeds the conventional production, the water in the upper reservoir is released and activates the turbine(s), thus producing the electricity that is injected into the network. For renewable electricity surpluses storage, the pumping power is critical as it has to cope with large volume of electricity in a relatively short time.

This technology has been used since 1920. In 2022, a worldwide capacity of 160 GW was available. It is divided into about 400 units with an average power of 0.4 GW. The largest station is located in the USA (Virginia), with a capacity of 3 GW and a storage capacity of 30 GWh.

In 1999, Japan had a unit in Okinawa using the sea as a lower reservoir (coastal PHS). It has been operating until 2016. This approach is interesting for the local control of electricity supply because 80% of the world's population lives near the coast.

An **hybrid system of wind turbine-pumped hydro** is operational in Germany at Gaildorf with four wind turbines with a power of 13.6 MW. At the foot of each, a tank of 40,000 m³ allows to store the water of a lower basin with 200 m level difference when the price of electricity is low or in case of surplus production. The generation power of the turbines is 16 MW.

Studies have also been made in Germany and other countries for the use of **abandoned coal mines** as a lower reservoir and a surface lake. The advantages are high dropping heights (up to 1,000 m), allowing turbines of several hundred megawatts to be used, as well as heat recovery, water reaching up to 40°C. In Australia, a project is considering a combination of a 150 MW photovoltaic park and the use of unused gold mines as a lower reservoir to produce up to 330 MW of electricity. Another option has been demonstrated by the Scottish Gravitricity: using weights (tens or hundreds of tonnes) falling from a tower or in shaft.

Although the theoretical potential is important for some countries, the achievements are limited to the optimal adequacy between the maximum possible power (depending on the height of the fall and the size of the reservoirs), the geological constraints and the high investments.

2.3.3.2 Compressed air energy storage

In the compressed air energy storage (CAES) approach, electricity is used, when surpluses are generated or prices are low, to compress air and store it in natural cavities. If needed, this air is released in a turbine producing electricity with a yield of 40–70% (Figure 2.13).

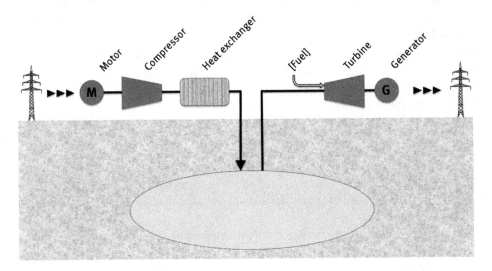

Figure 2.13: Compressed air storage principle.

Diabatic-CAES: In 2022, there are still only two important operational installations working in diabatic mode: after compression, the air, at a temperature of the order of 600°C, is cooled and injected into a cavity. If electricity is required, the compressed air is released and mixed with natural gas to power a turbine. The first plant was located in Huntorf, Germany, in operation since 1978 with a cavity of 310,000 m³ at 600 m depth. The turbine can provide 320 MW of power for 2 h. In the USA, the McIntosh plant, which has been in operation since 1991, can supply 226 MW (110 MW initially) for 26 h with 538,000 m³ of storage at 450 m depth. In 1991 a 110 MW storage was built in McIntosh, Alabama. It could run for 26 h with an efficiency of 25%. In 2021, China has connected to the grid a 100 MW/40 MWh CAES located in Changzhou. The US company General Compression developed a prototype using a 2 MW turbine in Texas. Compressed air stored can supply 1.6 MW for about 15 h. In California, Hydrostor is building a 480 MW unit with an operational target by 2026. Other projects or developments are conducted in the US, Japan, South-Africa.

    Adiabatic-CAES: With this technology, the heat of compression is recovered and stored in ceramics. When compressed air is released, it is passed through a heat exchanger to recover the stored heat and supplies the turbine without any use of fuel. A yield of above 70% is expected, if the few ongoing projects (The German ADELE project – 2013-2016 – has investigated different options and the consequences on cost, efficiency and the potential locations in Germany and the Netherlands) are realised, which does not seem to be the case.

Liquid air energy storage: Called sometimes cryogenic energy storage, it is still in the state of demonstrator. The air is compressed and then liquefied and stored at –169°C. The heat of compression is also recovered and used to vaporise the liquid air which actuates a turbine. Among the advantages, one can mention the high energetic capacity of liquefied air (214 Wh/kg, more than 10 times that of compressed air at 100 bar) and storage in tanks on the ground. The installation of the English company Highview Power Storage was a pilot unit near the Heathrow Airport with a power of 350 kW, connected to the electricity grid from 2011 to 2014. Another unit (50 M/300 MWh) is under construction near Manchester.

2.3.3.3 Battery storage

Storage in batteries offers, compared to previous solutions, modulated power and capacity that can vary from residential (a few kWh) to regulation of power plants (several hundred MWh). In addition, the storage capacity can be increased simply by adding additional batteries.

If battery technologies can take different approaches, the most spread are lead or lithium-ion batteries or derivatives (lithium/phosphate). Lead batteries are still used despite their low-energy density (30–50 Wh/kg for conventional technology), given their low cost.

Leading battery technology is lithium-based. If R&D work started in 1912 (US chemist Gilbert Newton Lewis), it has been developed and commercialised by Sony in 1991. Depending on the application, different chemistries are used: $LiFePO_4$, $LiMn_2O_4$, $LiNiCoAl_2$, $LiNiMnCoO_2$. Lithium batteries are using liquid electrolyte but solid-state batteries are ready for commercialisation in 2023 (Taiwanese company Prologium). However, other approaches without lithium are in the prototype or R&D phase.

Both types of batteries are sharing 50/50 the market for grid storage in 2022. R&D for lead batteries (nano-scale carbon, silicon mats. . .) allows improvements in terms of reduced weight, improved charge efficiency or reduction of lead content.

Li-ion technology will dominate mainly due to the constant fall of their price (from US $1,000/kWh in 2010 to about US $400/kWh in 2015 and US $132/kWh in 2021) following the increased global production capacity. The trend predicted in 2015 (Figure 2.15) has confirmed: it stands by about US $120/kWh in 2021. However, there was a then price increase, accentuated by the war in Ukraine in 2022, but it was also due to the high demand, COVID pandemic effect on workers in mines and treatment plants. . . . Another aspect is the availability of sufficient raw material to meet the demand.

Lithium-ion **gigafactories** are expanding worldwide. If we consider plants above 50 MWh capacity, there were five operating plants or projects in the US and nine in Europe in 2022. The most important cell manufacturers are still Asian companies

(China, Japan and South Korea) with a market share in 2021 of 32.5% for CATL (Contemporary Amperex Technology)/China, 21.5% for LG Energy Solution/South Korea, 14.7% for Panasonic/Japan, 6.9% for BYD/China, 5.4% for Samsung/South Korea and 5.1% for SK innovation/South Korea.

Worldwide, the number of **storage facilities** is rapidly increasing in terms of total installed power (12.4 GW in 2021, increasing to an expected 27 GW in 2025 and 70 GW in 2030) but also in terms of installations above 100 MW power. In 2021, America and China represented 70% of installed capacity. The US energy supplier Vistra has installed in 2022 in California the largest battery storage facility (Figure 2.14) with an available power of 400 MW (1,600 MWh). It will be supplemented by an extension adding 350 MW (1,400 MWh). In Hawaii, targeting 100% renewable electricity by 2045, the grid will be stabilised by a 185 MW (565 MWh) storage supplied by Tesla. In Germany, RWE will optimise power generation with two locations using batteries (Lingen/45 MW and Werne/75 MW) installed in containers, and operational in 2022. In Scotland, the Canadian AMP Energy is installing 800 MW (1.6 GWh) storage capacity to regulate Scottish wind farms electricity (25 GW of offshore wind will be added by 2032).

Another track is also to give **a second life for EV batteries** by integrating them in storage containers. EV batteries have usually 10 (end of first life) or 15 (end of second life) years' service life. Another criterion is the decrease of original capacity. Depending on manufacturers, when it decreases by 10 to 30% it is considered as having reached end-of-life. By 2030 there will be about 26 GWh available. If considered as still operational, these batteries can be used for stationary storage (domestic or utilities) where they can be operated until reaching about 40% of original capacity.

In Germany, Audi and the energy supplier RWE have installed in 2021 in Herdecke 60 decommissioned EV batteries (having "only" 80% of original capacity) being able to store 4.5 GWh. In Japan, Nissan EV batteries are used by East Japan Railway Company for emergency power. In Spain, Nissan and ENEL are experimenting in 2022 in Melilla a backup generator using 48 used and 30 new EV batteries giving a power 4 MW (1.7 MWh).

Battery recycling will need to make a big increase in terms of facilities and capacities. Batteries are manufactured with efficiency and cost criteria in mind. This means that most components are welded making the disassembly complex and time-consuming. The complex internal structure makes also separation of the different basic elements more difficult. However, it is even more important to separate the critical metals as the demand could exceed the extraction. In Norway, Northvolt and Hydro have started a recycling plant having an annual capacity of 12,000 ton of battery packs (covering Norway's needs). They claim that up to 95% of metals can be recovered. In the US Li-Cycle's plant in Arizona has an initial yearly capacity of about 10,000 ton. Another plant in Alabama has a capacity of 20,000 ton. Both units processing's capacity could increase up to 65,000 ton in 2023. China has the most recycling facilities worldwide (100,000 ton capacity).

Batteries for stationary application: lead-acid or lithium-ion?

Lead-acid battery is the oldest and well-established electricity storage technology. Although lithium-based batteries are dominating for electric vehicles, lead batteries are still predominant for some applications (12 V batteries or uninterruptible power supply) but also worth considering for stationary storage. The main advantage is the cost per kilowatt hours: in 2021 about US $120 for lithium batteries versus US $100 for lead batteries but with an important potential to reduce it. New developments to improve performance, life time and decrease manufacturing cost are conducted by universities or manufacturers: like bipolar plates, new materials (expanders to accelerate recharge) and process optimisation. The mid-long-term target is to reach a cost of US $35/kWh.

Recycling of lead batteries is a well-established technology and a profitable business. Up to 99% of end-of-life lead batteries are recycled and lead can be easily and efficiently recovered.

Figure 2.14: Storage park with capacity of 200 MW in Texas (Source: Vistra).

Behind-the-meter batteries: This concept covers domestic photovoltaic coupling with household batteries (up to 10 kWh of storage). This approach allows local use of the electricity produced to reduce the load on the networks and, depending on the price of electricity, reduces costs. In Germany, following a tax incentive campaign, more than 400,000 units were installed in 2022.

Sodium–sulphur batteries (NaS) commercialised by the Japanese company NGK insulators are operating at high temperatures (at least 300°C) and offer a storage cost per kWh lower than the one of Li-ion batteries, for example. A 20-foot container represents a power of 800 kW and a capacity of 4,800 kWh. The total installed global power is of the order of a few hundred megawatts, especially in Japan. The Futamata Wind Development Co., Ltd. uses 17 units of 34 MW to stabilise electricity supplied by a 51 MW wind

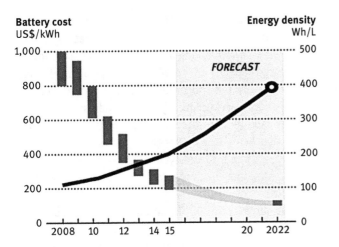

Figure 2.15: Evolution of prices and energy density of Li-ion batteries (US Department of Energy-projection in 2015).

farm. In 2018, batteries of 108 MW power (648 MWh) were installed in Abu Dhabi, UAE. Other units are used in Mongolia, Taiwan or the Netherlands (BASF Company).

Other electrochemical systems also allow the storage of electricity. The circulating batteries or **redox-flow batteries** (Figure 2.16) use two reservoirs containing the electrolytes circulated in the core of the cell composed of two chambers separated by a membrane.

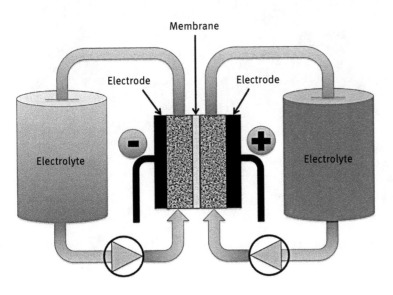

Figure 2.16: Principle of a circulating (redox-flow) battery.

The electrolytes generally used are vanadium salts. This technology allows storing large quantities of electricity compared to other types of batteries. Between 2005 and 2008, an experiment was carried out in Japan involving a 30 MW wind farm combined with circulating batteries with a storage capacity of 6 MW. Other wind farms in Australia, Scotland use these batteries to smooth out short fluctuations in production. The German company Vanadis Power is commercialising different models: the 2 MWh storage capacity/750 kW peak power module is integrated into a container, with the dimensions of the complete installation being 13.6 × 6.9 × 4.1 m. It has an AC–AC efficiency of about 65% and a lifetime of 20 years (>10,000 cycles). The smaller redox-flow batteries (90–120 kW) fit into a 20 ft container. In China, Vanadis Power is building a storage capacity of 200 MW (800 MWh).

The redox-flow batteries offer an alternative to lithium batteries in terms of lifetime, number of cycles and cost per kWh. The downside is a lower efficiency but compensated by the fairly easy increase in capacity by adding more tank storage. Many suppliers are active on this market (Sumitomo Electric/Japan, Australian Vanadium, Enerox/Austria, Dalian Rongke Power/China (Figure 2.17), UniEnergy Technologies/US, Vionx Energy/USA).

Figure 2.17: 500 kW/2,000 kWh redox-flow module (Dalian Rongke Power Co. Ltd.).

2.3.3.4 Thermal storage

In this approach, excess electricity is converted to heat (power-to-heat) and stored in water or other fluid tanks. If this can be applied to small or very large scale, but this conversion is irreversible. Existing plants will be described in the chapter power-to-heat (Chapter 6.2).

2.3.3.5 Electric vehicle batteries (vehicle-to-grid)

In many publications, batteries of electric vehicles are considered as short-term storage of electricity and possibly be used as an auxiliary source if necessary (Figure 2.18).

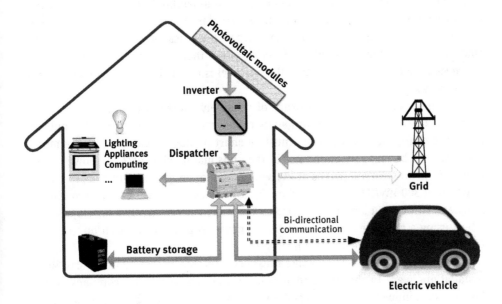

Figure 2.18: Storage and use of electric vehicle batteries.

While the approach seems elegant, it faces many limitations:

- the number of electric vehicles in circulation is still too low to have an impact on the management of electricity flows.
- the option for batteries to be a source for external uses is not always really provided by car manufacturers.
- who would accept to see the autonomy of his electric vehicle still reduced?
- up to what level batteries are discharged without hindering users?

However, large fleet of several million electric vehicles would offer a significant storage and regulation capacity: 1 million vehicles with a 30 kWh battery each and a 50% capacity utilisation rate would allow to store (and to recover) 15 GWh. In 2018, the German energy provider TenneT and Nissan launched a pilot project. Nissan's EV are equipped with the CHAdeMO connector which is bidirectional. The project was achieved in 2020 and allowed to validate the energy management *ChargePilot*. Other pilot projects were also conducted (Porsche and the German operator TransnetBW in Germany, Hyundai with the Dutch We Drive Solar in the Netherlands and Cradle Berlin in Germany). In July 2022, the German government funded a program to define the framework for V2G application.

2.3.3.6 Flywheel

A flywheel consists of a rotating mass (synthetic material, steel or other metal) mechanically connected to a motor/generator. The assembly is generally placed in a chamber with reduced pressure to minimise friction. When one wants to store electricity, the motor is started, driving the flywheel (electrical energy to kinetic energy). Once the motor is not powered, the steering wheel then continues its rotation. When it is necessary to recover electricity, the flywheel is coupled to the motor which then operates as a generator.

Flywheels are suitable for storage of relatively short duration (from a few minutes to a few tens of minutes) with an extremely short response time (a few millisecond). The charge–discharge cycles must be short because, depending on the model, the losses can be up to 5% per hour. For the units marketed in 2022, the power is generally of the order of a few tens of kW with a maximum of 2 MW. In 2015, in Japan a flywheel of 300 kW/100 kWh of carbon fibre was developed with a superconducting magnetic suspension system. The French company Energiestro has developed a flywheel using a concrete rotor. The available power is planned to reach up to 200 kW with a maximum storage of 50 kWh.

Existing facilities use parks consisting of several flywheels. By 2016, nearly 80% of installations were in the USA and 10% in Europe. The American company Beacon Power has equipped numerous installations mainly for frequency regulation: 20 MW in New York (200 units of 100 kW each) and in Pennsylvania in 2014 with the same power. In Munich, in 2015, SWM (Stadtwerk München) has installed 28 Stornetic units of 22 kW each capable of supplying 100 kWh. The 28 flywheels are installed in a 40-foot container. The US company Amber Kinetics 8 kW/32 kWh flywheel has been installed in many countries: in 2017, 16 units at the West Boylston Municipal Lighting Plant, Massachusetts, in 2018 in Hawaii and Lhasa, Tibet and in 2019 in Taiwan.

2.3.3.7 Supercapacitors

It is basically a capacitor with a modified structure enabling it to store a large amount of electricity with a very short charging time and to release it almost without loss and if necessary in a very short time. The supercapacitors are also characterised by an important number of charge–discharge cycles of the order of several hundreds of thousands. The main limitation to their use is the high cost for high power. They are mainly used to stabilise grids for short periods of time (a few seconds). In comparison, the Tesla Li-ion module 2170 measures 21 mm in diameter and 70 mm in length for a weight of 66 g and can store 21.3 Wh. To store the same amount of electricity, it would take about five supercapacitors from Maxwell of 3 V and 3,000 F (unit storage capacity of 3.75 Wh at 3 V DC) weighing 2.6 kg. Skeleton/Estonia with plants in Germany or Nawa/France with a plant in 2023 are using graphene (skeleton, Figure 2.19) or carbon nanotubes (Nawa) which provide high energy density and fast charge time compared to conventional supercapacitors.

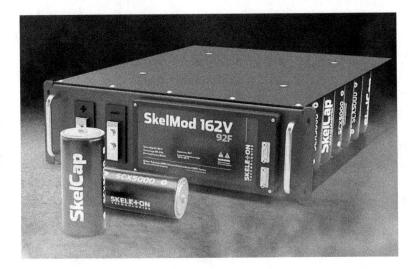

Figure 2.19: Supercapacitor modules and 162 V pack of 538 kW/335 Wh (Source: Skeleton Technologies GmbH).

The main limitation of supercapacitors is their high cost for high power (US $2,000–10,000/kWh in 2022 for electric vehicle application). However, they have a very high number of cycles of charge/discharge (up to 1 million) and a very short charging/discharging time (ms).

2.3.3.8 Superconducting magnetic energy storage
Electricity is stored as a magnetic field produced by electricity circulating in a coil maintained at very low temperatures, requiring complex cryogenic system using hydrogen or liquid helium. If the response time is very short and the efficiency is high (>95%), the cost is also due mainly to the equipment necessary to maintain the low temperatures.

2.3.3.9 Rail vehicles
The American company Advanced Rail Energy Storage proposes to use wagon-weighted electric locomotives on a sloping track to store electricity when the locomotive rises up

the slope. For release, the locomotive and wagons descend and electricity is recovered through braking energy, the engine acting as generator. After a demonstrator with a 5.7-tonne vehicle and a 268 m track, a more powerful project in the state of Nevada was to provide (for a total of 8 km track and a total of 32 locomotives of 272 tonnes each) a capacity of 1.5 MW with a yield of 80%, i.e., a total power of 50 MW/12.5 MWh. The cost would have been US $55 million with an installation lifetime of 30–40 years. In 2022, it is still a project with 210 cars on 10 parallel tracks weighing 75,000 tonnes.

2.3.3.10 Underwater storage

The possibility to store onsite or close offshore excess electricity opens the way to new technologies. The German Institute Fraunhofer has experimented a variation of pumped hydro using submerged concrete spheres as lower and upper reservoirs. Filling with seawater takes place under pressure and generates electricity. When the price of electricity is low or the electricity is in surplus, the sphere is emptied to start a new cycle. At 500 m depth, the energy produced reaches 1.4 kWh/m^3. This technology could be combined with offshore wind farms. The Dutch company Ocean Grazer has developed a concept with a storage unit on the seabed. Water in a concrete tank (10 MWh capacity) is pumped in a flexible reservoir when excess electricity is available and stored under pressure. When power is needed, the water is released through a turbine. A prototype was tested in 2021.

2.3.4 Comparison of available electricity storage solutions

Considering the different operational technologies around the world, pumped hydro dominated and accounted still for more than 96% of global storage capacity in 2022 (about 160 GW). In comparison, batteries for network stabilisation accounted for 27 GW (7.0 GW in the USA).

The different technologies are at different levels of development, which explains either their cost (low or high) or their degree of implementation (Figure 2.20).

2.3.5 Characteristics of electricity storage technologies

Each electricity storage technology has characteristics that condition the use: cost, reactivity, storage capacity etc. Figure 2.21 illustrates the different approaches (Table 2.6).

The main issue of the solutions presented is the global limited storage capacity over time due to their principle: when the batteries are charged or when the upper reservoir of the PHS or the compressed air storage caverns are filled, it is no longer

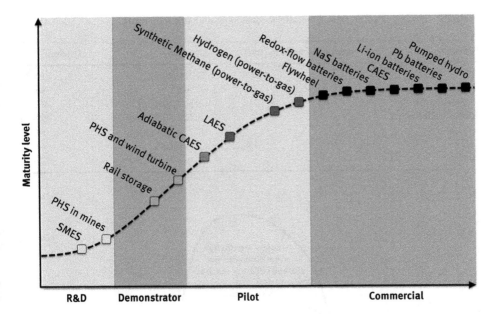

Figure 2.20: Maturity of different storage technologies.

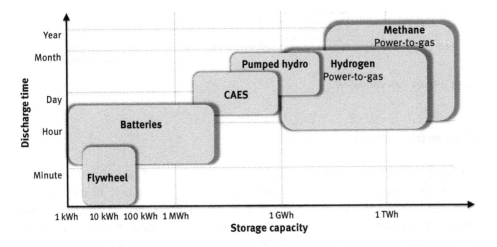

Figure 2.21: Electrical storage technologies.

possible to store the surplus provided by the wind or photovoltaic (Figure 2.22). An extension of capacity, if possible, requires new investments.

It is therefore necessary to find an approach to lift this limitation and use a technology that would offer virtually unlimited storage.

Table 2.6: Comparison of the main characteristics of storage solutions – 2021 data.

Technology	World power 2021 (GW)	Lifetime	Efficiency (%)
PHS	160	30–60 years	70–85
Thermal	3.1	30 years	80–90
Li-ion batteries	27	Up to 10,000 cycles	85–95
Redox-flow batteries		Up to 14,000 cycles	60–85
NaS batteries		Up to 4,500 cycles	70–90
CAES	1.6	20–40 years	40–70
Flywheels		Up to 100,000 cycles	70–95

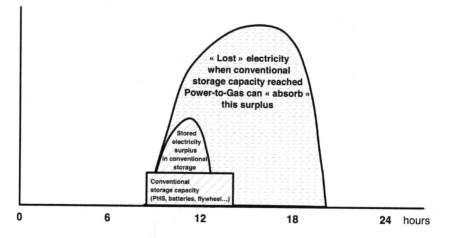

Figure 2.22: Intermittent storage.

2.3.6 Electricity storage requirements

Worldwide, the 195 GW of storage available in 2021 accounted for only about 2.6% of the electricity generation capacity. The installed global capacities, all technologies combined, are (still?) very small compared to the consumptions, which limit the capacity of regulation. The generation of large quantities of electricity of renewable origin in the coming decades as well as the expected surpluses will require volumes/power of storage that are not commensurate with current capacities.

The German DIW Institute (Deutsches Institut für Wirtschaftsforschung) published in 2021 a study [2] showing that a complete renewable energy supply would be possible by 2030. Electricity generation will be dominant: 223 GW for onshore wind, 820 GW for offshore wind and 1,126 GW solar PV. Storage will be provided by batteries (30 GW) or Pumped Hydro Storage and hydrogen (indirect storage through 60 GW of electrolysers).

The growth of electricity from renewable sources is related to storage

The development of the electricity market and the trend towards decentralisation go hand in hand with storage solutions at all levels: residential, tertiary and industrial. For each of these sectors there are appropriate technologies. However, at the level of the global network, solutions allowing storage of large volumes will prove necessary.

The actual overall available storage capacity (mainly PHS) related to consumption shows that it can at best smooth out peaks of local demand. The increase in renewable electricity production with inevitable fluctuations leading to production sometimes higher than demand will require storage capacities that are not comparable with what is currently available. The various limitations of "classical" solutions (cost, limited storage capacity, sometimes little or no extension possible) leave room for another technology that could potentially meet these challenges: the **power-to-gas** which is a technology breakthrough in storage.

References

[1]   Global Wind Energy Council GWEC, Global Wind Report 2022.
[2]   DIW Weekly Report 29+30/2021, 100% Renewable Energy for Germany: Coordinated Expansion Planning Needed.

3 Principle of power-to-gas

When electricity production from renewable sources is greater than consumption, what can be done with this surplus electricity, which can be considered as "free"?

This is the challenge that power-to-gas (P2G) technology wants to address.

P2G is the use of surplus renewable electricity to produce a gas that can be stored or used directly.

What are the conditions and technologies involved?
– Have significant renewable electricity generation
– Have at certain times a production higher than consumption
– Directly produce gas using this electricity
– Store, use or process this gas

The "transformation" of electricity into gas is possible, thanks to the **electrolysis** which allows the dissociation of water in hydrogen and oxygen.

An (relatively) old technology

In 1889, the Dane Poul la Cour (1846–1908) converted a traditional mill into a windmill (Figure 3.1) and associated it with a dynamo and a developed regulator, the cratostat. The objective then was to be able to store this electricity to be independent of the meteorological conditions. After rejecting batteries due to their high cost, he turned to electrolysis (Figure 3.2), producing hydrogen and oxygen stored in separate tanks with a volume of 12 m³. The seven electrolysers were supplied by Italian professor Pompeo Garuti. Production could reach 1,000 L of hydrogen per hour. Hydrogen was then used directly for lighting the mill and a nearby school. He then realised that the hydrogen/oxygen mixture could be used for autogenous welding. His attempt in 1902 to modify an engine to run it with hydrogen and thus have electricity continuously did not succeed.

A museum located in Vejen, Denmark occupying the original mill (which has since undergone many modifications) presents the work of this pioneer.

All elements of the P2G concept were already gathered:
– wind turbine for power generation
– electrolyser for conversion to hydrogen
– storage for future use

3.1 Basic layout

Excess renewable electricity is used by an **electrolyser**. This produces oxygen and hydrogen which can be used in many ways (Figure 3.3):
– Directly for the industry, for example (petrochemistry, chemistry, electronics etc.)
– Injection into the natural gas network
– Production of methane (CH_4) by methanation, eventually injected into the natural gas network

https://doi.org/10.1515/9783110781892-004

Figure 3.1: Windmill transformed into a wind turbine by Poul la Cour (Poul la Cour Museum/Vejen).

Hydrogen or methane produced by mixing with natural gas can then be stored in reservoirs for natural gas (tanks or natural caverns).

3.2 Hydrogen

Hydrogen was identified as an element by Henry Cavendish in 1766. It was Lavoisier who gave its name in 1788 (hydrogenium) from the Greek roots "hydro" (water) and "genes" (from). In 1800, the English Nicholson and Carlisle used for the first time the electrolysis of water to produce it. It was liquefied by James Dewar in 1898.

Hydrogen is the simplest element: the atom consists of a nucleus formed by a single proton around which orbits an electron.

Hydrogen is not an energy source, but an **energy vector**. If it is the most widespread element in the universe, there is practically no free form on earth.

Figure 3.2: Electrolysers used by Poul la Cour (Poul la Cour Museum, Vejen).

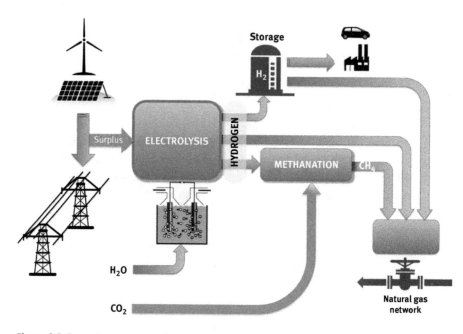

Figure 3.3: Power-to-gas concept diagram.

3.2.1 Properties

Hydrogen is the lightest element (molecular weight = 2.016 g for 22.4 L under normal conditions of temperature and pressure), which is an advantage for balloons or dirigibles, but a disadvantage for its transportation and its storage. It is a colourless, odourless, non-toxic gas.

3.2.1.1 A high energy density
When compared with other fuels, hydrogen contains for the same mass the largest amount of energy (Figure 3.4).

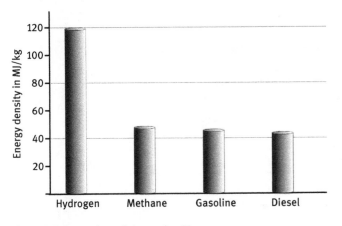

Figure 3.4: Comparison of energy densities.

However, this high energy density of hydrogen corresponds to large volumes due to its low density.

3.2.1.2 Low density
In gaseous state, its density is 0.0899 with respect to air (1 Nm^3 of hydrogen weighs only 89.9 g compared to 1204 g/Nm^3 for air or 651 g/Nm^3 for gas natural). Liquefied at a temperature of 252.76 °C (20.39 K), its density is only 70.79 g/L.

These two characteristics explain the large volumes required for storage or special isolation for liquid hydrogen tanks. In addition, liquefaction of hydrogen requires large quantities of energy (up to 40–50% of its energy content depending on the capacity of the production unit) (Table 3.1).

Table 3.1: Energies required for hydrogen compression or liquefaction.

Initial energy content	Compression energy	Liquefaction energy/capacity	Final energy content (MJ/kg)
100% to 142 MJ/kg		80 MJ/kg @ 100 kg/day	62
		50 MJ/kg @ 10,000 kg/day	92
	14 MJ/kg @ 200 bar		128
	22 MJ/kg @ 700 bar		120

3.2.1.3 Large storage volumes needed

For the same amount of final energy (in this case 250 kWh), the volumes required to store gasoline, natural gas or hydrogen are shown in Table 3.2 and Figure 3.5.

Table 3.2: Comparison of volumes required to store the same amount of energy.

	Gasoline	Natural gas	Compressed hydrogen (CGH$_2$)	Liquid hydrogen (LH$_2$)	
		200 bar	200 bar	700 bar	
Fuel	20 kg	18 kg	7 kg	7 kg	7 kg
	28 L	128 L	500 L	194 L	98 L
Tank	6 kg	70 kg	80 kg	120 kg	140 kg

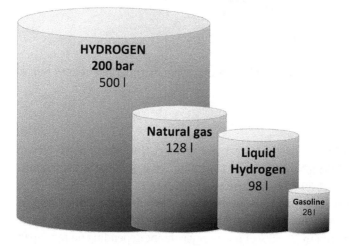

Figure 3.5: Comparison of volumes required to store the same amount of energy (not to scale).

Storage of hydrogen in reservoirs is certainly possible, but for very large volumes (thousands or millions of cubic metre), this solution is impractical.

Liquid hydrogen tanks for vehicles

Liquid hydrogen was tested to power fuel cell or thermal engine vehicles. Beginning 1990 saw ICE engine with hydrogen fuel experimented: passenger cars from Mazda in Japan (rotary Wankel engine and 60 miles range) or BMW in Germany ("Hydrogen 7" with 260 HP engine and 200 km range), busses from MAN, Ballard and Daimler Benz . . . . The tank of the model BMW Hydrogen 7 (IC engine) presented late 2000s containing 8 kg of liquid hydrogen and weighed 168 kg, a ratio of 1:21. The volume of the tank was about 300 L.

3.2.2 Security

Since the fire of the Hindenburg airship in 1937, the collective unconscious classifies hydrogen as dangerous. The public's negative perception of safety should not be underestimated for its widespread use.

It is certainly not without risk, as do all the other fuels (gasoline, diesel, natural gas etc.). Each, however, has its own specificities. Hydrogen has the following characteristics:

- the low minimum ignition energy (20 µJ, compared to 290 µJ for natural gas)
- in the event of a leak, hydrogen ignites more frequently than other gases
- the small size of the molecule allows it to diffuse through materials and to weaken them
- the flammability range is between 4% and 75%
- the self-ignition temperature is high (585 °C)

These characteristics do not make it an innocuous gas, but show that hydrogen can be used (largely in the industry) taking into account its specificities.

"City gas" and hydrogen

This gas (coal gas or town gas), obtained by pyrolysis of coal, has been used in many countries for decades, in the 1950s in the USA and in 1970 in Britain, among others by households for heating and cooking.

This gas, which varied in composition from country to country, consisted mainly of hydrogen (up to 60%), methane (up to 55%) and CO (up to 10%).

3.2.3 Industrial production

World consumption was estimated as 90 million tonnes in 2021 (about 1,015 billion m^3), produced mainly (96%) from non-renewable sources: natural gas (58%), coal (18%) or as a by-product of the chemical industry (20%) like chlorine production by electrolysis. Each tonne produced from natural gas generates 11 tonnes of CO_2. Less than 4% is produced by electrolysis, often from electricity produced by large dams (Egypt, Peru and Zimbabwe).

Hydrocarbons reforming

Reforming or steam reforming involves reacting hydrogenated compounds (hydrocarbons such as natural gas, gasoline, diesel or alcohols such as methanol and ethanol) or coal with steam or oxygen according to the reaction (for hydrocarbons):

$$C_nH_m + nH_2O \Rightarrow nCO + (n + m/2)\ H_2$$

About 95% of hydrogen is currently used in petroleum (desulphurisation) and chemical or steel industry (ammonia production or methanol), and the rest is commercially available (*merchant hydrogen*).

Hydrogen colour: The media, governments and some experts have classified the produced hydrogen according to a colour code: "green" for renewable energy, "blue" or "grey" from natural gas (with or without carbon capture), "turquoise" from pyrolysis, "pink" from nuclear electricity and even "white" for natural sources of hydrogen.

3.2.3.1 Hydrogen and electricity from renewable sources

Currently, the only method of storing large quantities of surplus renewable electricity is the production of hydrogen by electrolysis. This hydrogen opens the way, directly or indirectly, to many uses in industry, transportation and energy in all sectors of the economy.

3.2.4 "Carbon-free" industry and energy through carbon capture and storage

The carbon capture and storage (CCS) or carbon capture utilisation and storage (CCUS) technologies are considered as a way to decarbonise hydrogen production from natural gas or coal or some processes (steel or cement, for example) or the energy sector (coal or gas power plants). Storage usually takes place in adequate geological formations (deep saline) or at the bottom of the sea.

In Norway, the CO_2 storage on the Sleipner field started in 1996 is the first worldwide commercial project. Located about 250 km from the coast, the captured CO_2 is transported by ship and injected at a depth of 900 m. In 2021, 23 Mt CO_2 were stored. Norway's GHG emissions in 2020 were 49.3 Mt CO_2 equivalent (24.6 Mt from oil, gas extraction, industry and mining and 14.7 Mt from transportation).

In Iceland, Climeworks is capturing CO_2 from the air and injecting it in the underground. Capture capacity was 4,000 tonne/year in 2022.

Starting 2024, the Dutch Yara Company will capture CO_2 from its ammonia and fertiliser plant. Logistics will be organised by the Norwegian Norther Lights. The CO_2 will then be liquefied and transported by boat to Oygarden, Norway. After a temporary storage, it will be injected 100 km from the coat at a depth of 2.6 km. In 2025, up to 800,000 tonne/year will be treated. As a comparison, Netherlands has emitted 178 million tonnes CO_2 in 2021.

3.2.4.1 CO_2 emissions versus storage capacity

In 2021 about 30 facilities worldwide were in activity capturing and storing 40 million tonnes of CO_2. However, this has to be related to the estimated CO_2 emissions that were about 36.3 billion tonnes (36.3 Gt). Even if the very optimistic figures of 7.5 Gt of CO_2 captured and stored projected by the IEA in the report *"Net Zero by 2050: A Roadmap for the Global Energy Sector"* [1] can be achieved, it would still represent a small fraction of the emissions.

3.2.4.2 Unknown costs of storage

The CO_2 concentration in the exhausts produced by the industry is very variable and it has to be separated from other exhaust gases. The higher the concentration, the lower the capture costs. A 2021 estimate leaded to US \$35–150/tonne. The viability of this technology will depend on the CO_2 market price that should be high enough to cover the costs.

The other main limitations are the capture and transportation network to build, the geologically long-term safe storage areas to find, the acceptance by the population and the eventual risks of leaks like in 1986 where a lake in Cameroon released brutally 100,000–300,000 tonnes leading to the death of more than 1,700 people. However, the potential of CCS/CCUS seems to be overestimated being based on extrapolations of the existing but limited operations does not take into account the development of a true CO_2 free economy.

Reference

[1]   IEA. Net Zero by 2050, A Roadmap for the Energy Sector, October 2021.

4 Electrolysis

Discovered in 1800 by the English Nicholson and Carlisle, continuous electrolysis, which comes from Greek words (ἤλεκτρον [electron] "amber" and λύσις [lýsis] "dissolution"), allowed the production of hydrogen in large volumes. The reforming of natural gas and hydrocarbons or the gasification of coal then supplanted it, except in countries where electricity is abundant (e.g. Canada).

4.1 Basic principle

The key element of the power-to-gas chain is the **electrolyser** that allows the production of hydrogen from surplus renewable electricity. The electrolysers used can be grouped into two main families: alkaline and *proton exchange membrane* (PEM), whose characteristics will be described in detail. Other technologies are under evaluation or in development such as high-temperature electrolysis.

4.2 Chemical reactions

The decomposition of water by electricity makes it possible to separate hydrogen and oxygen according to the reaction:

$$2\,H_2O\,(1) + \text{energy} \Rightarrow 2\,H_2(g) + O_2(g) \tag{4.1}$$

The energy to be supplied for the dissociation of water consists of electrical and thermal energies:

$$2\,H_2O + [237.2\,\text{kJ/mol of electricity} + 48.6\,\text{kJ/mol of heat}] \Rightarrow 2\,H_2 + O_2 \tag{4.2}$$

An electrolysis unit comprises an electrolyte and two electrodes separated by a membrane or diaphragm (Figure 4.1).

In an electrochemical cell (Figure 4.2), water is decomposed if a certain voltage (critical voltage) is applied between the two electrodes.

At equilibrium, there is always a partial dissociation of water into H^+ and OH^- ions. In alkaline medium, OH^- ions dominate.

At the anode (positive electrode), OH^- ions are decomposed according to the reaction whose standard potential E_0 is −0.4 V:

$$2\,OH^- \Leftrightarrow H_2O + {}^1\!/_2\,O_2 + 2\,\bar{e} \tag{4.3}$$

https://doi.org/10.1515/9783110781892-005

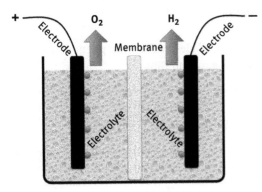

Figure 4.1: Principle of water electrolysis.

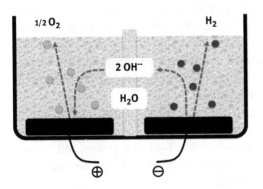

Figure 4.2: Principle of an electrochemical cell (alkaline electrolysis).

At the cathode (negative electrode), the reaction is carried out with a standard potential E_0 of -0.827 V:

$$2\,H_2O + 2\bar{e} \Leftrightarrow H_2 + 2\,OH^-$$ (4.4)

The minimum potential required for the water decomposition reaction is 1.227 V under normal temperature and pressure conditions. The standard potential E_0 of the oxidation-reducing reaction, expressed in volts, is measured with respect to a reference electrode.

Electrolysis in acid medium

In this medium, H^+ protons predominate.
Reaction at the anode:

$$H_2O \Leftrightarrow {}^1/_2\,O_2 + 2\,H^+ + 2\bar{e}$$ (4.5)

Reaction at the cathode:

$$2\,H^+ + 2\bar{e} \Leftrightarrow H_2$$ (4.6)

4.2.1 Calculation from thermodynamic data

Thermodynamic laws allowing the description of electrochemical phenomena have been established, among others, by Nernst, Faraday and Gibbs.

Faraday's laws relating to electrolysis

1. The mass of a substance altered at an electrode during electrolysis is directly proportional to the quantity of electricity Q transferred ($Q = I \times t$).
2. For a given quantity of electric charge Q, the mass of an elemental material altered at an electrode is directly proportional to the element's equivalent weight. The equivalent weight of a substance is equal to its molar mass divided by the change in oxidation state it undergoes upon electrolysis.

Electrical and eventually thermal energies are converted to "stored" chemical energy in hydrogen and oxygen products. The energy required for decomposition of water is the enthalpy of water formation ΔH (285.84 kJ/mol under normal conditions). Only the free energy ΔG, known as Gibbs free energy, is to be supplied to the electrodes in electrical form, the remainder being represented by the thermal energy as a function of the temperature and the entropy variation ΔS. The enthalpy change is given by the relation $\Delta G° = \Delta H° \, T\Delta S°$ (Gibbs–Helmholtz equation).

The *enthalpy* of a system corresponds to the total energy of this system. *Entropy* characterises the degree of disorder of a system.

Different equations govern the equilibrium of the H_2–O_2/H_2O system and make it possible to calculate the minimum potential required for electrolysis from thermodynamic data at constant pressure and temperature:

$$E^0_{cell} = \frac{-\Delta G^0}{nF}$$

where $\Delta G°$ is the free energy change of Gibbs, n is the number of electrons involved and F is the Faraday constant.

The Gibbs–Helmholtz equation $\Delta G° = \Delta H° - T\Delta S°$ allows to determine, from the data available for the elements involved in this reaction (the enthalpy of water formation is $\Delta H° = 285.84$ kJ/mol and the entropy $\Delta S°_{tot} = 0.163$ kJ/mol), the minimum theoretical voltage to start the electrolysis (*reversible voltage*) which is 1.227 V under a pressure of 1 bar and a temperature of 298 K (25 °C).

Without the addition or production of thermal energy (adiabatic conditions), the minimum isothermal decomposition voltage of water (thermoneutral voltage), which can also be calculated from the thermodynamic data, is 1.48 V under normal conditions (1 bar and 298 K) (Table 4.1).

The temperature and applied voltage affect the type of reaction and the hydrogen production (Figure 4.3).

The reversible voltage U_{rev} at a given temperature and pressure is defined by the Nernst equation:

$$U_{rev} = U_0 + \frac{RT}{2F} * \ln \frac{P_{H_2} P_{O_2}^{1/2}}{P_{H_2O}}$$

where P_{H_2}, P_{O_2} and P_{H_2O} being the operating pressures of the electrolyser and U_0 the reversible voltage under normal conditions: Figure 4.4 shows the reversible voltage as a function of pressure at different temperatures.

Table 4.1: Minimum theoretical voltage for electrolysis.

Conditions	Reversible voltage	Thermoneutral voltage
$P = 1$ bar/$T = 298$ K	1.227 V	1.48 V

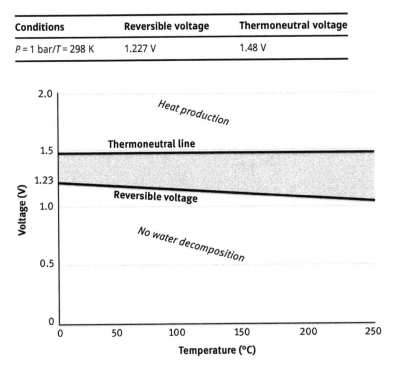

Figure 4.3: Influence of temperature and voltage on electrolysis parameters (ideal cell).

Depending on the temperature at which the electrolysis is carried out (Figure 4.5), three zones are distinguished: one where hydrogen cannot be produced, one where the reaction is endothermic (need to supply heat) and a third where the reaction is exothermic (production of heat to be evacuated). Along the thermoneutral line no heat input or cooling is required.

The actual operating voltages vary between about 1.7 and 2.0 V. This corresponds to a yield of the order of 75–85%.

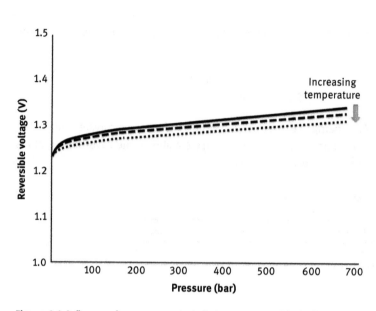

Figure 4.4: Influence of pressure on electrolysis parameters (ideal cell).

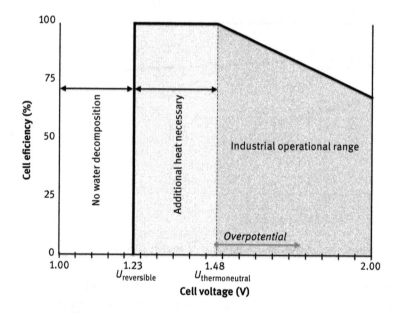

Figure 4.5: Efficiency of a cell as a function of voltage.

4.2.2 Operating voltage – current density

If voltage is the important factor in producing hydrogen under optimum conditions, the current applied plays a role on the amount of hydrogen produced (Figure 4.6).

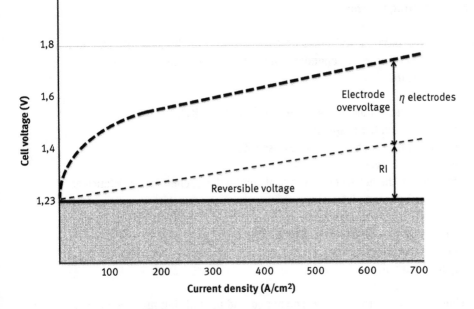

Figure 4.6: Actual operating voltage according to current density.

This current density (A/cm^2) is a function of the voltage and internal resistance of the cell or stack. As a first approximation, this overall resistance is the sum of the resistances of the various elements (anode, electrolyte, gaseous layer, membrane, gaseous layer, electrolyte and cathode):

$$R_{total} = R_{electrode} + R_{electrolyte} + R_{bubbles} + R_{membrane}$$

At the theoretical voltages (reversible or thermoneutral), the polarisation voltages η (overvoltages) due to the electrical resistances of the various components must be added:

$$E = E_0 + \eta_{electrodes} + \eta_{electrolyte} + \eta_{bubbles} + \eta_{membrane}$$

which can also be expressed in another form:

$$E = E_0 + \eta_{electrodes} + RI$$

where R is the sum of resistances in Ohm/cm^2 due to the electrolyte, the bubble formation and their migration as well as those due to the membrane and I the current density (A/cm^2).

The actual operating voltage is the result of the polarisation voltages and is a function of the current density. For low current densities, the theoretical efficiency increases because the activation energy and the ohmic overvoltages are low. However, low current densities result in lower hydrogen production.

4.2.3 Operating parameters

Many factors have an influence on the operation of the electrolyser and hence the production of hydrogen. Depending on the technology used they are:
- temperature,
- pressure,
- electrical resistance of the electrolyte (conductivity),
- electrolyte quality (impurities),
- concentration of the electrolyte (viscosity),
- electrolyte flow,
- material of the electrodes (electrical conductivity, chemical resistance etc.),
- distance between electrodes,
- electrode size and alignment,
- gas bubbles on the surface of electrodes,
- membrane material and
- type of current.

During electrolysis, the water is consumed and an addition must be made in order to have the same concentration of electrolyte for electrolysis in an alkaline medium. Moreover, the circulation of the water/electrolyte solution or of the water must make it possible to evacuate the gas bubbles (H_2 and O_2) formed on the surface of the electrode and optionally to homogenise the concentration of the solution.

4.2.4 Cell yield

The electrical efficiency of a cell can be calculated according to the following formula:

$$\eta_{electrical}/\% = \frac{100*(U_{anode} - U_{cathode})}{U_{cell}}$$

Which heating value to consider?

The higher heating value (HHV) includes all energy released by a reaction between initial and final state at the same temperature (usually 25 °C). It should therefore be used for efficiency calculation.

Another expression is Faraday's performance:

$$\eta_{\text{Faraday}} = \frac{\Delta G}{\Delta G + \text{losses}}$$

It can also be expressed as a function of thermal efficiency:

$$\eta_{\text{thermal}} = \frac{\Delta H}{\Delta G + \text{losses}}$$

Use of lower heating value?

Depending on the final application, the HHV (285 kJ/mol) or LHV (241.8 kJ/mol) of hydrogen could be considered.

For electrolysis using water in liquid form, we consider the HHV to calculate the yield.

If the heat of condensation of water is not considered the lower heating value could be used. For a cell voltage of 1.8 V, for example, the respective yields are:

-   1.48/1.8 = 0.82 (82%) considering the higher calorific value
-   1.23/1.8 = 0.69 (69%) considering the lower calorific value

Since the system is not ideal, other losses and consumption of auxiliary equipment (pumps, control system etc.) must also be taken into account.

The actual efficiency is also a function of the type and size of the electrolyser: the higher the efficiency, the higher the yield which can reach 85% or more for the 2017 electrolyser.

4.2.5 Water dissociation energy

How much electricity is needed to dissociate water? If the thermodynamic data give the theoretical values, parasitic phenomena (e.g. resistance of the various components) as well as the auxiliary equipment require higher energy.

The theoretical yield of an electrolyser corresponds to HHV of hydrogen, which in normal conditions (298 K and 1 bar) requires **3.54 kWh/Nm^3** or 39.4 kWh/kg (assuming that all the heat of decomposition of the water is recovered and the final temperature of the water is equal to its initial temperature).

Physical parameters such as material structure, internal layout, ageing of certain components and auxiliary equipment (pumps, compressors etc.) mean that the actual energy required for dissociation is higher than the theoretical value.

For a 75% efficiency, the energy to supply will be theoretically 4.7 kWh/Nm^3 or 52 kWh/kg.

4.2.6 Water consumption

The theoretical quantity of water required for electrolysis is given by the reaction:

$$H_2O \Rightarrow H_2 + {}^1\!/_2\,O_2 \tag{4.7}$$

For 1 mol of hydrogen produced (22.4 L) the quantity of water required will be 18 g or 18 cm^3 (Table 4.2). Based on stoichiometry, 1 kg of hydrogen and 9 L of water are required.

Table 4.2: Amount of water required as a function of the volume of hydrogen.

Hydrogen volume	Water volume
100 m^3	80,400 L
1,000 m^3	804 m^3
1,000,000 m^3	804,000 m^3

These figures show an often neglected parameter, which is the high water consumption, if we consider surplus electricity from renewable sources in the coming decades.

If an energy requirement of about 5 kWh/Nm^3 of hydrogen is assumed, an annual surplus of 1 TWh corresponding to the production of 200 million m^3 of hydrogen would require at least 160 million m^3 of water.

For comparison, the yearly water consumption of Berlin (3.7 million inhabitants) is about 215 million m^3 in 2021 and Germany's household water consumption in 2021 was about 4.77 billion m^3. Replacing the hydrogen produced from fossil fuels (about 90 Mt in 2021) by electrolysis would require about 800 million m^3 water.

4.2.6.1 Two main families of electrolysers

Although it was originally an acid solution that made it possible to demonstrate the phenomenon, it was then alkaline electrolysis that was first used in industry, allowing chemicals that are easier to use and relatively less aggressive for the materials.

Another technology was later developed, based on the use of specific membranes where protons (H$^+$) were circulating (PEM). Following the development of the high-temperature fuel cell (solid oxide fuel cell – SOFC), a new technology based on it was developed. The latest technology, anion exchange membrane (AEM), is now being implemented: unlike alkaline electrolyser, the half-cell layout allows a better separation of hydrogen and oxygen generation and a higher hydrogen purity.

4.3 Alkaline electrolyser

This type, the oldest used industrially, is based on an alkaline electrolyte. It benefits from a long experience resulting in a reliable and competitive technology.

4.3.1 History and industrial development

In April 1800, English scientists Nicholson and Carlisle observed the decomposition of water by electricity into two gases. They used a Volta battery discovered the previous year. The German Johann Wilhelm Ritter quantifies in the same year the production of hydrogen and oxygen by electrolysis.

Early in the twentieth century, more than 400 industrial electrolysers were in operation. They were using electricity of hydraulic origin. The hydrogen produced was used for the synthesis of ammonia (NH_3) and the manufacturing of fertilisers. As early as 1927, high-power electrolysers (1 MW and higher) were available from the Norwegian company Norsk Hydro, now NEL Hydrogen. From 1927, production sites used up to 150 U, i.e. 150 MW. Figure 4.7 shows electrolysers installed in Rjukan, Norway.

Figure 4.7: Alkaline electrolysers in 1927 (NEL Hydrogen).

4.3.2 Operating parameters

4.3.2.1 Components
An electrolysis cell (Figure 4.8) comprises an alkaline solution, two electrodes and a membrane (or diaphragm) allowing the passage of OH⁻ ions.

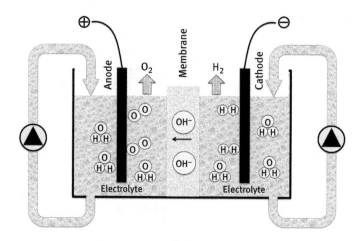

Figure 4.8: Basic cell of an alkaline electrolyser.

The most commonly used electrolyte is potassium hydroxide (KOH), which is more conductive and less aggressive than sodium hydroxide, at a concentration between 25% and 30%. Additives are used in order to increase the ionic activity or reduce the aggressiveness of the electrolyte.

The diaphragm is made of a porous material, which has the role of letting the OH⁻ ions circulate, separating the hydrogen from the oxygen formed. The materials used are either asbestos, micro-perforated fabric or nickel in sintered form. In the latter case, the diaphragm must not be in contact with the electrodes.

The electrodes (anode or cathode) are generally made of nickel or its alloys (Ni/Fe, Ni/Co/Zn, Ni/Mo etc.) with sometimes electrolytic or vapour deposition of other metals (Zn, Co, Fe, Pt etc.) to increase the reaction rate.

4.3.2.2 Effects of temperature and pressure
The Gibbs–Helmholtz equation involves temperature in the calculation of the operating voltage. It is a factor that influences performance. Generally, operating temperatures for alkaline electrolysers vary between 40 and 90 °C. Depending on the design of the stacks, electrolysers can supply hydrogen (and oxygen) at atmospheric pressure or under pressure up to 60 bar (Table 4.3).

Table 4.3: Comparison of atmospheric/high-pressure configurations.

Electrolyser type	Atmospheric pressure	High pressure
Advantages	Simple configuration Long industrial experience Reduced costs Easier control	Compact design at identical power Possibility to increase the number of cells
Disadvantages	Large size More complex gas drying More limited number of stacks	Higher costs Safety Reduced operating range at high pressure

The combined influence of these two parameters (*P* and *T*) on the output is illustrated in Figure 4.9. At equal current density, the pressure plays a negligible role, while the efficiency is improved if the temperature increases.

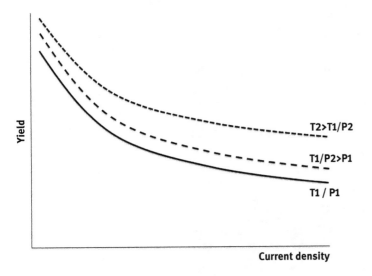

Figure 4.9: Schematic influence of temperature and pressure on performance.

4.3.2.3 Current type
The direct current (DC) used is generally constant. However, a current of variable amplitude about a mean value or a pulsed current may be used (Figure 4.10).

4.3.2.4 Loss of voltage
Electrolysis is a complex phenomenon due to the numerous interfaces involved and the reactions at these interfaces.

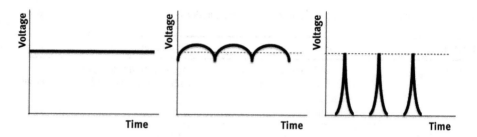

Figure 4.10: Current types (DC voltage–steady DC voltage waveform–short pulse DC voltage).

For the electrochemical part, the voltage losses at each interface and in the various components accumulate. Figure 4.11 gives an idea of these losses:
– losses at anode and cathode
– overvoltages at the surface of electrodes
– losses due to gas bubbles
– losses due to electrolyte
– loss due to membrane

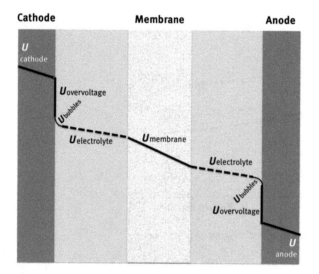

Figure 4.11: Voltage losses along the chain.

4.3.2.5 Gas contamination

The membrane is not completely gas-tight, which results in oxygen diffusion into the compartment where hydrogen is produced and vice versa. The gases produced also carry KOH and water to be separated. The purity of the hydrogen obtained is generally greater than 99.5%.

4.3.3 Structure of an alkaline electrolyser

The electrodes and membrane (or diaphragm) can be arranged (Figure 4.12) with or without spacing (zero gap).

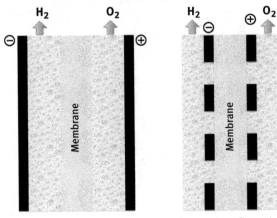

Zero-gap configuration

Figure 4.12: Layout of electrodes and membrane.

The layout of the cell groups can have two configurations: monopolar and bipolar.

In the monopolar stack (Figure 4.13), all anodes and cathodes are fed directly by the current source which supplies a voltage equal to that of a cell.

For the bipolar configuration (Figure 4.14), only the terminal electrodes are directly supplied with current with a voltage corresponding to the sum of the voltages of each cell.

The industrial electrolysers (Figure 4.15) have practically all adopted the bipolar structure, which allows a higher current density. Table 4.4 compares these two configurations of electrodes positioning.

4.3.4 Auxiliary equipment

The electrolyser is associated with different equipment (balance of plant, Figure 4.16) necessary for optimal operation:
- a water demineralisation and feeding unit (pumps),
- a unit for separating oxygen from hydrogen,
- optionally a unit for separating hydrogen from oxygen,
- drying units for both gases,
- a temperature control unit,
- DC power (voltage and current) and
- a control and management unit for the various parameters.

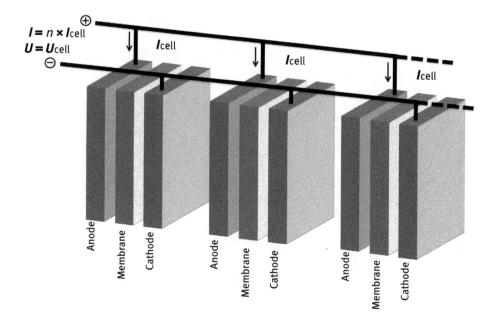

Figure 4.13: Monopolar design.

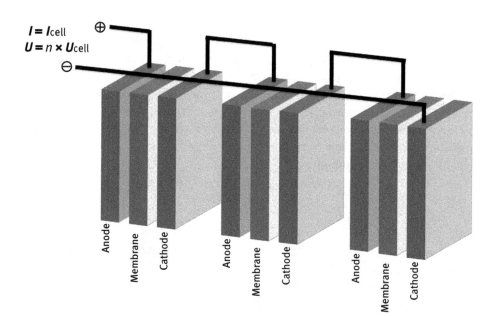

Figure 4.14: Bipolar design.

Figure 4.15: Alkaline electrolyser stack (NEL Hydrogen).

Table 4.4: Comparison of monopolar and bipolar configurations.

	Monopolar design	**Bipolar design**
Characteristics	Low operation voltage (of one cell) High current density	Voltage equal to the sum of the voltage of each cell
Advantages	Lower investments Defective cells may be short-circuited	Compacity Higher yield
Disadvantages	Large size Low potential for yield improvement	Higher cost If a cell is not working, dismantling the rack is needed

4.3.5 Industrial equipment

4.3.5.1 Power of electrolysers

This power is calculated taking into account the current density, the supply voltage and the number of cells. It allows to optimise the design according to the volume of hydrogen to be produced.

The most important existing industrial plants operate continuously, often using hydropower and can produce up to 30,000 Nm^3/h of hydrogen, used for the production

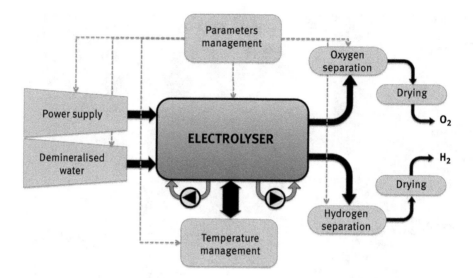

Figure 4.16: Auxiliary equipment (BOP – balance of plant).

of ammonia and fertilisers. In 2022, the Japanese supplier AsahiKASEI presented a 10 MW zero-gap design alkaline electrolyser (peak production: 2,000 Nm^3 H2/h).

4.3.5.2 Start-up time

In the case of cold start, alkaline electrolysers require up to 20 min to be operational. For power-to-gas, the start-up should ideally be instantaneous in order not to lose electricity when there is surplus production. To bypass this limitation, isolating the electrolyser to maintain the temperature above 30 °C may allow a start in a few seconds.

4.3.5.3 Lifetime

For an alkaline electrolyser, the elements that wear are the membranes and the electrodes that must be regularly revised. Depending on the size of the installation, the type of operation (continuous or not) or other parameters, the lifetime of a unit varies between 50,000 and 90,000 h.

4.3.5.4 Suppliers

Only some high-power electrolyser suppliers that are compatible with the power-to-gas concept are listed.

The Norwegian company NEL Hydrogen (formerly Norsk Hydro) has been designing electrolysers since the 1920s. Model A 485 (Figure 4.17) with a power of 2.2 MW can produce up to 3,880 Nm^3/h of hydrogen under 1 bar at a purity of 99.9%.

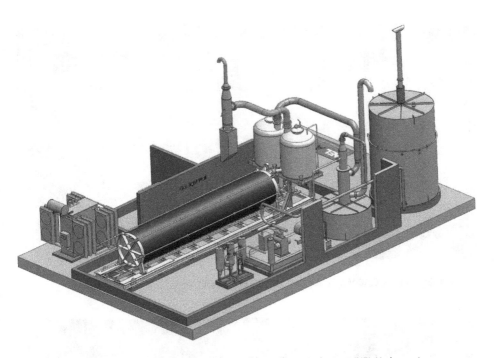

Figure 4.17: High-power A 485 alkaline electrolyser with auxiliary equipment (NEL Hydrogen).

The Japanese AsahiKASEI which started electrolysis in the 1920s is involved in the commercialisation of alkaline electrolyser (Aqualyzer/10 MW) and different projects including the European ALIGN-CCUS (CO_2 capture and use with green hydrogen to produce synthetic fuels). The AsahiKASEI atmospheric alkaline electrolyser Aqualyzer has a power of 10 MW (Figure 4.18).

Green Hydrogen Systems/Denmark is commercialising a pressurised alkaline electrolyser (Figure 4.19), the Hyprovide A90. The 40-foot container with a power of 0.9 MW (180 Nm^3/h-390 kg/day of hydrogen at 35 bar) includes all auxiliaries. It provides a ramp-up from 0% to 100% of 5 s.

Other companies also offer alkaline electrolysers with a capacity greater than 1 MW, capable of producing several hundred cubic metres of hydrogen per hour, such as Sunfire in Germany (having acquired the Swiss company IHT), McPhy in France, Jingli, Longi or Tianjin Mainland Hydrogen Equipment in China.

4.3.5.5 The alkaline electrolyser, a proven technology

Alkaline electrolysers have a long experience in optimising performance (materials used, efficiency, lifetime and cost). The power range available (up to several megawatts) allows it to be used for the exploration phase of power-to-gas technology.

Figure 4.18: Asahi KASEI 10 MW Aqualyzer™ alkaline electrolyser.

Figure 4.19: Green Hydrogen Systems alkaline electrolyser Hyprovide.

4.4 PEM electrolyser

Following the development of PEM fuel cells for the US space programme, General Electric used the same technology to produce an electrolyser in 1966 (solid polymer electrolyser). The first high-power commercial models (100 kW, 20 Nm^3 of hydrogen per hour) were produced by the company ABB from 1987.

4.4.1 Principle

A membrane coated with a **catalyst** separates the anode from the cathode (Figure 4.20). The decomposition reactions of the water occur at the anode.

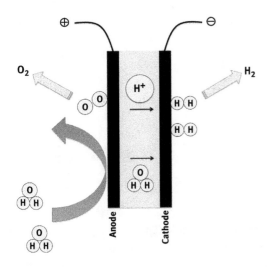

Figure 4.20: Principle of the PEM electrolyser.

The water is decomposed at the anode:

$$2\,H_2O \Rightarrow O_2 + 4\,H^+ + 4\,\bar{e} \tag{4.8}$$

After passing through the membrane, the protons H$^+$ react with the cathode to form a hydrogen atom:

$$4\,H^+ + 4\,\bar{e} \Rightarrow 2\,H_2 \tag{4.9}$$

The core of the PEM electrolyser is the PEM membrane. This reaction is very slow and only the use of catalysts accelerates its speed.

4.4.2 PEM electrolyser structure

The central component is the MEA (membrane–electrode assembly) (Figure 4.21). It comprises a membrane with a thickness of the order of 100–300 µm, coated on each side with catalyst generally based on precious metal (Precious Metal Group, often platinum, iridium and ruthenium) as well as the electrodes (or current collectors).

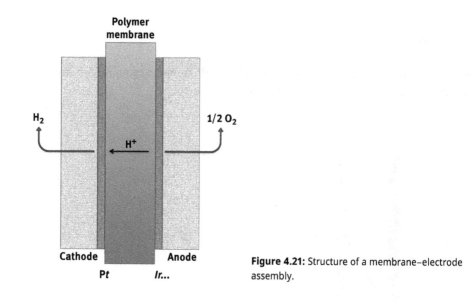

Figure 4.21: Structure of a membrane–electrode assembly.

The membrane should have the following characteristics:
- high ionic conductivity in S/cm to allow proton circulation,
- low permeability to gases, especially hydrogen and oxygen in order to avoid their diffusion in the other compartment,
- very good resistance to chemicals,
- very good thermal and mechanical stability and
- high permeability to water.

The **membrane** is a material allowing the passage of protons H^+. The oldest is the Nafion® that was developed in the 1960s by the American company Du Pont de Nemours. It is commercialised by Chemours or Merck (Sigma-Aldrich) and available in the form of powder or sheets of different thicknesses.

Nafion® (Figure 4.22) is a perfluorosulphonic acid/polytetrafluoroethylene stabilised composite copolymer in the acid form (H^+). Sulphonic groups SO_3^- favour the conductivity of protons.

Other membranes have been developed either around this family of copolymers like sulfonated polyether ether ketone or able to work at higher temperatures (above 100 °C). Those are based on the family of polybenzimidazole doped with an acid

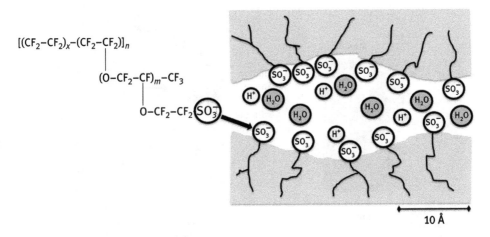

$$[(CF_2-CF_2)_x-(CF_2-CF_2)]_n$$

$$(O-CF_2-CF)_m-CF_3$$

$$O-CF_2-CF_2\,SO_3^-$$

10 Å

Figure 4.22: Chemical structure of Nafion®.

(phosphoric or sulphuric acid) or pyridines. BASF's Celtec membranes are available for high-temperature applications (120–180 °C). The cost of the membranes is generally high. In 2022, a 0.61 × 2.50 m sheet of Nafion 117 (18 μm thick) costed US $3,880.

This membrane is coated with catalyst (e.g. platinum for the cathode and iridium or ruthenium for the anode) in the form of very fine particles in order to have a large contact surface.

A **current collector** in contact with the catalyst allows the flow of current and the evacuation of the gases produced (GDL or gas diffusion layer). It must be porous, good electrical conductor, resistant to corrosion, allows the anode to pass water to the catalytic sites and evacuates the oxygen produced. The materials used are titanium in the form of sintered powder or in the form of carbon tissues at the cathode.

The **bipolar plates** serve as electrodes and circulators for water and for the product gas (hydrogen at the cathode and oxygen at the anode). They have channels where these elements circulate. They also separate the cells from each other (Figure 4.23).

An elementary cells assembly (MEA, current distributors, bipolar plate) forms a stack (Figure 4.24). Commercial electrolysers are composed of several tens of stacks (Figure 4.25).

4.4.3 Balance of plant

As for alkaline electrolysers, they are all auxiliaries around a cell or a stack (see Figure 4.16).

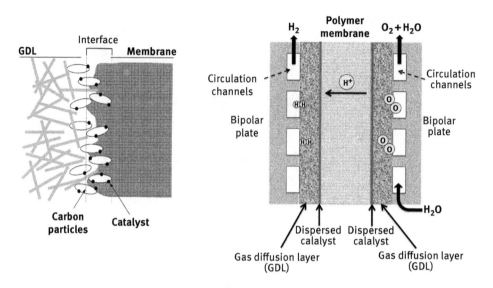

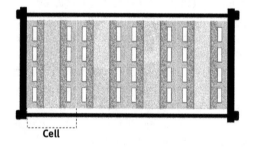

Figure 4.23: Diagram of the components of a PEM electrolytic cell.

Figure 4.24: Assembling of elementary cells (stack).

4.4.4 Influence of the catalyst

4.4.4.1 Catalyst loading

The catalyst acts as an accelerator for the decomposition of water. It is therefore necessary to optimise the concentration (expressed in milligram per square centimetre or gram per square metre of membrane), its structure (the finest particle size possible) and its distribution on the membrane (homogeneous dispersion).

The high cost of the catalysts results in a shift towards increasingly low loads from a few milligrams per square centimetre to less than 1 mg/cm^2, the compromise being excellent catalytic activity and catalyst stability over time (little deterioration in performance). Even if these charges appear to be small, extended to several square metre, they represent a significant cost.

Figure 4.25: Stack for PEM electrolyser (Siemens).

4.4.4.2 Precious metals group

The metal used as catalyst at the cathode is platinum, optionally combined with other metals. This metal is produced mainly in South Africa (more than 70% of world production), Russia and Canada. Its price is very variable (Figure 4.26 in US$ per troy ounce, 31.1 g) and strongly influences the price of the MEA. By way of comparison, over the past 10 years, the price of gold has risen to a maximum of US $2050 per ounce with an average of US $1,200.

For the anode, it is the ruthenium (more than 90% coming from South Africa) and especially the iridium (85% from South Africa, as co-product from platinum and palladium extraction) which is more stable or their oxides which serve as catalyst.

In PEM electrolysers, catalysts are important contributors to the stack cost, itself representing up to 60% of overall cost.

Replacing platinum?

By its high and variable price, platinum contributes to the higher cost of PEM electrolysers. For many years, numerous studies have been conducted in order to find an alternative, however, until 2022 without success. While some metals have shown positive laboratory results, their catalytic properties decline rapidly over time (a few hundred hours at best). The work in the industry is oriented rather towards a reduction of the charge of precious metals while maintaining the catalytic activity over a long period.

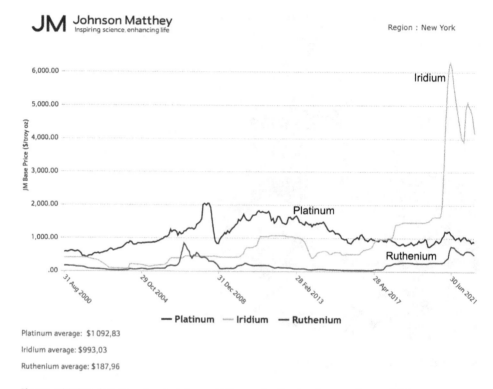

Region : New York

Platinum average: $1 092,83

Iridium average: $993,03

Ruthenium average: $187,96

Figure 4.26: Trends in the price of platinum, iridium and ruthenium between August 2000 and September 2022 (Johnson Matthey).

4.4.5 Operating parameters

As with the alkaline electrolyser, factors such as temperature and pressure affect the performance of the PEM electrolyser.

4.4.5.1 Temperature

For a given voltage, hydrogen production increases with temperature. It is important, however, not to exceed a certain threshold which depends on the nature of the membrane so as not to degrade it. Generally, the temperature remains between 70 and 90 °C for the low-voltage operating PEM electrolyser.

4.4.5.2 Pressure

PEM-type electrolysers can directly supply hydrogen under high pressures (over 100 bar). In 2016, pressures of up to 350 bar were reached with prototypes.

4.4.5.3 Current density

PEM-type electrolysers can be used at higher current densities than alkaline electrolysers, thus permitting higher hydrogen production. The densities can reach several A/cm^2 depending on the parameters, including the thickness of the membrane.

4.4.5.4 Membrane area

This parameter determines the volume of hydrogen production by influencing the applied current density per unit area of the membrane (in Ampere per square centimeters or Ampere per square meters). This current density must be optimised to achieve both high hydrogen production and also to have a maximum lifetime of the membrane and catalysts.

4.4.5.5 Efficiency of the electrolyser

As for alkaline electrolysis where the thermodynamic parameters of water decomposition are identical, the theoretical minimum energy of decomposition along the thermoneutral curve is 3.54 kWh/Nm^3 of hydrogen if one considers the HHV.

The required energy also depends on the maximum production capacity of the electrolyser. Generally, it varies between 5 and 8 kWh/Nm^3 depending on the capacity of the units. At atmospheric pressure, hydrogen is produced with a purity of more than 99.995% (class 5).

4.4.5.6 Lifetime

Degrading elements are the membrane and the catalysts (the particles combine reducing the catalytic efficiency). In 2022, the reported service life exceeds 50,000 h in continuous operation. However, compared to alkaline electrolyser more factors have a negative influence on life time: higher pressure, temperature, current density and variable load in operation. To overcome those, "overdesign" (thick membrane, higher catalyst load etc.) is used but this increases the overall cost.

4.4.6 Industrial equipment

Only some suppliers of high-power PEM electrolysers, compatible with the power-to-gas concept, are presented.

The modular Siemens Silyzer 300 (Figure 4.27 and Table 4.5) with a power of 10 MW can produce hydrogen under a pressure of 35 bar.

The Silyzer 300 electrolyser was evaluated in Austria as part of the European project H2Future before commercialisation in 2018.

The high-power PEM electrolyser stack of ITM Power (Figure 4.28) can produce up to 285 Nm^3/h of hydrogen under 40 bar. It measures $1.00 \times 0.80 \times 0.55$ m.

Figure 4.27: 10 MW electrolyser (Siemens).

Table 4.5: Characteristics of the Silyzer 300.

Silyzer 300	Specifications
Power	10 MW
Size with auxiliary equipment (in metres)	15 × 7 × 3.7
Weight of module	2.1 ton
Starting time	<1 million
Hydrogen purity	99.5–99.999%
Nominal production	Up to 2,000 Nm^3/h
Water consumption	10 L/Nm^3 H$_2$
Lifetime	>80,000 h

The PEM electrolyser market has seen concentrations (NEL/overtook Proton On-Site, GTT/France Areva H2GEN and Cummins/USA Hydrogenics Corp.) and many new entries (Ohmium/USA, Green Hydrogen Systems etc.).

China is also more and more present in the PEM electrolyser market. (The Sungrow Company, active in the PV business, has brought a 250 kW PEM electrolyser in 2021.) Joint ventures with western companies are also signed (Cummins/USA and Sinopec for a manufacturing plant).

Figure 4.28: ITM Power 2 MW electrolyser stack (Photo: author at Hannover Fair).

Among suppliers with high power electrolysers:
- NEL MC500 2.5 MW – 500 Nm^3/h
- H-TEC/Germany ME450 – 1 MW – 210 Nm^3/h
- ITM Power/UK HGAS3SP – 2 MW – 370 Nm^3/h
- Cummins/USA HyLYZER-1000 – 2.5 MW – 1,000 Nm^3/h
- GTT Elogen/France ELYTE 260 1.3 MW – 260 Nm^3/h

4.4.7 Water treatment and consumption

The starting element for electrolysis is water. The cells of the PEM electrolyser are susceptible to numerous contaminations. For this reason, water must meet certain criteria, such as conductivity and impurity concentration.

The power-to-gas installation in Mainz, Germany, uses a PEM-type electrolyser and treats the water in several stages (Figure 4.29) to provide demineralised water.

4.4.7.1 PEM electrolysis, strong potential for improvement
The more complex structure than the alkaline electrolysis and some specific components (membrane and catalysts) make the PEM electrolyser still expensive. This technology is

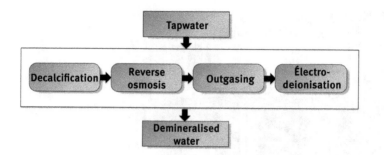

Figure 4.29: Preliminary water treatment.

relatively recent compared to alkaline electrolysis, but has a high potential to improving performance in terms of overall efficiency and cost reduction.

Due to the high demand for PEM electrolysers "Gigafactories" (at least 1,000 MW of yearly capacity) are operational, under construction or planned. The objective is to have a supply capacity to meet the market. The UK company ITM Power had in 2021 a first factory with a capacity of 1,000 MW/year running and is building a second one.

4.5 High-temperature electrolyser

This technology (SOEC – solid oxide electrolyser cell) shown in Figure 4.30 aims to achieve high efficiency with lower electricity consumption than other families of electrolysers. The decomposition reaction of water takes place at high temperatures (500–900 °C) and requires an external heat source which can be obtained from an industrial process. High temperatures require specific materials such as ceramics for electrodes and electrolyte.

At high temperature, the water is vaporised and decomposed at the cathode according to the reaction:

$$2\,H_2O + 4\,\bar{e} \Rightarrow 2\,H_2 + 2\,O^{2-} \tag{4.10}$$

The anions O^{2-} pass through the membrane and form oxygen at the anode:

$$2\,O^{2-} \Rightarrow O_2 + 4\,\bar{e} \tag{4.11}$$

The thermodynamic conditions for this reaction are energetically more favourable than low-temperature electrolysis. The Gibbs ΔG energy of the reaction goes from 237 kJ/mol at room temperature to 183 kJ/mol at 900 °C while the molar enthalpy ΔH (total energy required) varies little (Figure 4.31). Some of the energy can be supplied by an external heat source ($T\Delta S$).

The structure (Figure 4.32) is composed of a **solid electrolyte** (YSZ-stabilized mixed oxide of zirconium and yttrium) and electrodes, a mixed oxide of lanthanum, strontium and manganese (LSM) for the anode and a YSZ/nickel mixture for the cathode. The

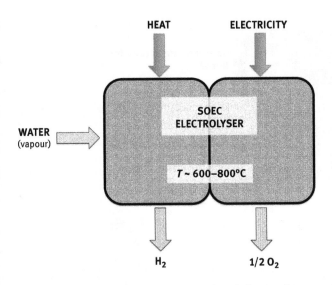

Figure 4.30: Principle of high-temperature electrolysis.

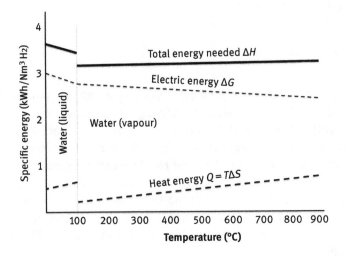

Figure 4.31: Energy balance as a function of temperature.

electrodes have a porous structure and water decomposition reactions take place on their surface. The O^{2-} ions circulate in the electrolyte using the crystalline defects.

High-temperature electrolysis allows a lower voltage than alkaline or PEM and a high current density (up to 12 A/cm^2 in laboratory). The main challenge is the development of the ceramics for electrolyte and the interface electrolyte/electrodes: degradation and delamination at working temperature should be minimised to reach durability and keep efficiency during lifetime. The developments are also oriented towards a reversible

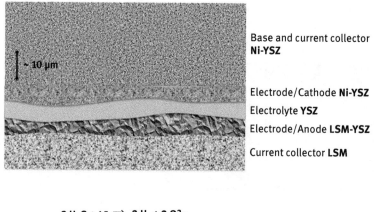

Base and current collector **Ni-YSZ**

Electrode/Cathode **Ni-YSZ**
Electrolyte **YSZ**
Electrode/Anode **LSM-YSZ**

Current collector **LSM**

$$2\,H_2O + 4\bar{e} \Rightarrow 2\,H_2 + 2\,O^{2-}$$

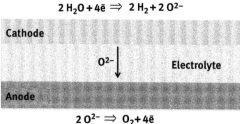

$$2\,O^{2-} \Rightarrow O_2 + 4\bar{e}$$

Figure 4.32: Structure of a cell and reactions.

solid oxide cell system where the electrolyser (solid oxide electrolytic cell) can also be used as a fuel cell (solid oxide fuel cell) using hydrogen, natural gas or methane.

Sunfire developed in 2016 a prototype of 200 kW operating at a temperature of 850 °C and supplying hydrogen under a pressure of 30 bar. The external heat input from another industrial process reduces the electrical requirements by about 16% (electrical efficiency >90%). The Sunfire HyLink (Figure 4.33) basic module has a power of 2.68 MW that can be modulated between 50 and 100% and it can produce up to 750 Nm³/h. Sunfire is involved in many programmes: MULTIPLY in a refinery in Rotterdam, DEMO4GRID in Switzerland for a refuelling station, P2X SOLUTIONS in Finland for synthetic fuel production etc.

The Danish company Haldor Topsoe is building a manufacturing plant in Denmark, operational in 2024, with a yearly capacity of 500 MW expandable to 5 GW. The 100 MW electrolyser has a capacity of 32,000 Nm³/h. Bloom Energy/USA, supplier of SOFC, has developed a reversible 270 kW demonstrator SOEC electrolyser. The Dutch company VoltaChem started in 2020 development to improve the performance of SOEC.

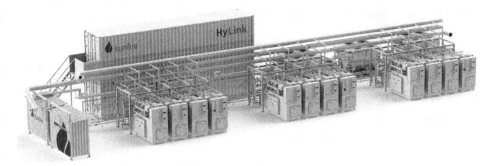

Figure 4.33: Sunfire HyLink (Sunfire).

Nuclear power plant and SOEC

Heat from a nuclear reactor could be used to supply superheated steam at 500 °C which would feed a high-temperature electrolyser. The French EDF and AREVA NP evaluated in 2016 hydrogen production from a nuclear reactor of 600 MW_{th}. The electricity consumption of the electrolyser would be of the order of 3.2 kWh/Nm^3 but would require an additional infrastructure. None of the numerous studies (International Atomic Energy Association) to use nuclear reactors have yet had an industrial follow-up.

SOEC technology offers a better overall efficiency of electrolysis than other technologies, and reversibility makes possible to convert it quickly into a fuel cell with an electrical efficiency of 55%. The high temperature avoids the use of catalyst like platinum for PEM electrolyser. With the German company Sunfire, Haldor Topsoe/Denmark, and Japan's Toshiba (production planned for 2025) and USA's Fuel Cell Energy are also active in this sector. Research is mainly focused on improving the degradation of cell performance.

4.6 Other technologies

4.6.1 Microbial electrolysis

Waste water treatment relies on a continuous flow. A microbial electrolysis cell can remove organics from waste water and produce hydrogen. Microbes are used as biocatalysts. The basic cell consists in two chambers and an ion exchange membrane. Single chamber cells have also been evaluated. At the anode surface, microbes create a biofilm that can convert the chemical energy in electrons. At the cathode, these electrons can produce hydrogen or/and methane. The global reaction at the anode is:

$$\text{Organic matter} \Rightarrow CO_2 + (CH_4) + H^+ + \bar{e}$$

At the cathode, where platinum can be used:

$$2\,H^+ + 2\bar{e} \Rightarrow H_2$$

Microbial electrolysis can be combined with anaerobic digester to increase the methane production yield. Complex pollutants can also be removed from waste water. However, although connected with great potential, industrial feasibility has not yet been proven.

4.6.2 High-frequency electrolysis

This technology uses the dissociation of water at very high temperatures (>3,000 K). These temperatures are reached by a thermal plasma using a torch initially fuelled by air which is gradually replaced by water (liquid or vapour).

The German company Graforce Hydro GmbH has developed an electrolyser using a plasma which would achieve a very high efficiency and a low cost of the hydrogen produced (€2–4/kg claimed in 2022). No water is used. In 2022, its Plasmalyzer® plants can produce up to 6,500 Nm³/h of hydrogen using methane, biomethane, biomass or waste water. By-product is elemental carbon (3 kg for 1 kg of hydrogen when using methane or biomethane).

4.6.3 Photoelectrolysis (photolysis)

In nature, the production of hydrogen by biophotolysis takes place in two stages: photosynthesis by light irradiation of green algae or cyanobacteria and production of hydrogen catalysed by enzymes.

Industrial photocatalysis is based on the direct decomposition of water into hydrogen and oxygen under the effect of light. This approach promises a lower cost of production than electrolysis.

The phenomena involved can be:
- At the surface of an electrode (titanium oxide is often used)
- On the surface of a photocatalyst (e.g. rare earths) or a semiconductor
- Specialised microorganisms (algae or cyanobacteria)

Many problems remain to be solved. For biological photocatalysis, for example, the production of oxygen reduces the activity of the enzymes or the absorption of the greater part of the photons by the chlorophyll of the microalgae.

4.6.4 Solar hydrogen

This technology (photoelectrochemical) combines a photovoltaic panel and an electrolyser (PEM membrane) into a sandwich unit. The semiconductor element is in direct contact with the electrolyte, the oxidation–reduction reactions occurring at the surface. Laboratory studies have achieved a maximum yield of about 20%. The problems to be solved are, among others, corrosion, transportation of ions and gases in the cells.

4.7 Hydrogen purification

The hydrogen produced by the electrolyser must undergo a series of operations in order to bring it in line with subsequent uses, in particular, its purification to meet the criteria for the intended destination.

According to the applications envisaged, hydrogen must meet certain purity standards. The membranes or electrodes used are not completely impermeable to oxygen (crossover), of which small quantities (0.1–0.2%) are then mixed with hydrogen. The hydrogen produced also contains traces of water or KOH for alkaline electrolysis. The concentrations of contaminants depend on the operating parameters (temperature, pressure etc.).

4.7.1 Elimination of KOH

For alkaline electrolysis, the use of a scrubber eliminates the traces of electrolyte, cools the hydrogen and collects the water for reuse after treatment.

4.7.2 Oxygen removal

The technology to be adopted for removing entrained oxygen will depend on the degree of purity required and the volumes to be treated.

4.7.2.1 Palladium/silver membrane
These membranes (Figure 4.34) make it possible to extract the hydrogen molecules by dissociating into monoatomic hydrogen at the surface of the membrane Pd/Ag, Pd/Au/Cu or Pd/Cu and then diffusing where ionisation takes place followed by recombination into diatomic hydrogen after passing through this membrane. This process provides high-purity hydrogen (>99.999%), but the cost of membranes is high.

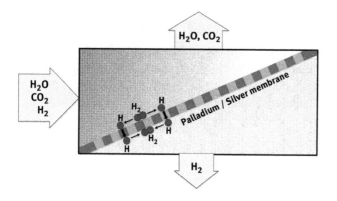

Figure 4.34: Membrane purification.

The main parameters are the pressure, the temperature (300–400 °C) and the thickness of the membrane. Regular purging of the surface of the membrane makes it possible to eliminate the molecules remaining on the surface.

Other membranes (ceramic, polymer) are studied, but their separating power is lower than that of palladium/silver.

4.7.2.2 Cryogenic process

The gas mixture to be purified is first treated to remove any compound that can solidify during the cryogenic treatment (water), then it is compressed and passes through a series of exchangers (where it is cooled by an external source and/or a series of detents by Joule–Thompson valve) and separators where the hydrocarbons condense, leaving an increasingly pure hydrogen.

4.7.2.3 Adsorption by pressure variations (pressure swing adsorption)

This process is based on the selective adsorption of all the components except hydrogen (due to its low molecular weight) by a substrate (activated carbon, molecular sieves etc.) and supplies high-purity hydrogen (up to 99.9999%). In general, the system (Figure 4.35) consists of a battery of several units, one part in operation (adsorption under pressure from 10 to 40 bar) and the other in regeneration (the collected adsorbed gases can be used to reheat the reformer). This process also makes it possible to process large volumes (units of more than 100,000 m^3/h).

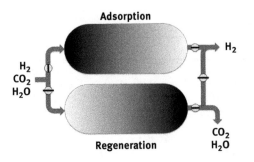

Adsorption

H₂

H₂
CO₂
H₂O

CO₂
H₂O

Regeneration

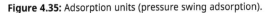

Figure 4.35: Adsorption units (pressure swing adsorption).

4.7.2.4 Use of selective catalyst

Oxygen is converted into water in a unit (DeOxo) using a catalyst (palladium on alumina carrier for the Puristar RO-20 from the BASF company).

4.7.3 Dehydration

The drying of hydrogen can be carried out by units equipped with molecular sieves at a temperature of −60 °C (dew point). The regeneration of the absorbent is done thermally. Water-resistant silica gels or glycol can also be used.

4.8 Technology comparison

The operational technologies over a wide range of power are alkaline and PEM. High-temperature electrolysis (SOEC) is still in a development phase.

4.8.1 Characteristics of electrolysers

Each technology offers characteristics that condition its use in a given field or for a specific application (Table 4.6).

PEM-type electrolysis has the advantage of avoiding the chemically aggressive potash solution and of having a wider operating range. It accepts a higher current density. However, it requires a specific membrane of high cost.

Alkaline electrolysis, on the other hand, benefits from decades of experience and optimisation of performance and costs.

SOEC electrolysers promise higher efficiency but have to show their durability in the long term due to the thermal stresses to which they are subjected (Figure 4.36).

Table 4.6: Comparison of alkaline, PEM and SOEC electrolysers.

Property	Alkaline	PEM	SOEC	AEM
Electrolyte	KOH	Membrane	Solid state	Membrane
Charge carrier	OH^-	H^+	O^{2-}	OH^-
Operating temperature	40–90 °C	Up to 100 °C	600–800 °C	50–60 °C
Hydrogen pressure	Up to 30 bar	Up to 200 bar	Up to 30 bar	Up to 30 bar
Electrodes	Ni/Fe	Pt, Ir, Ru	Mixed oxides	Ni or Ni alloys
Power variation range	20–100%	0–100%	0–100%	NA
Electrical efficiency (kWh/Nm^3)	4.5–5.0	4.5–9.0	About 4.0	
Current density	<1 A/cm^2	5–8 A/cm^2	Many A/cm^2	<2 A/cm^2
Rated life	8–15 years	4–7 years	>10 years	NA

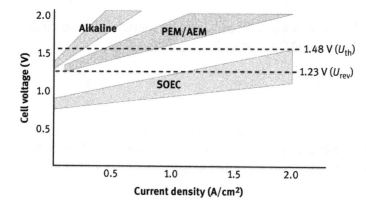

Figure 4.36: Comparison of the electrical parameters of the different electrolyser technologies.

4.8.2 Alkaline exchange membrane (AEM) electrolyser: the best compromise?

Alkaline electrolysers use KOH which is corrosive and can form carbonate deposits by reacting with CO_2 from the air. The PEM type uses a proton exchange membrane and precious metals as catalysts that increase the cost.

Research has led to the development of an alkaline membrane (AEM) based on polysulphones, thus eliminating the need of a liquid electrolyte, the OH^- ions, passing through the membrane.

The structure (Figure 4.37) is similar to the PEM electrolyser: an anion conductive membrane with diffusion layer/electrodes on both sides. Reactions at the electrode are

$$\text{at the cathode}: 2\,H_2O + 2\bar{e} \Rightarrow H_2 + 2OH^-$$

$$\text{at the anode}: 2\,OH^- \Rightarrow 2\bar{e} + H_2O + \frac{1}{2}O_2$$

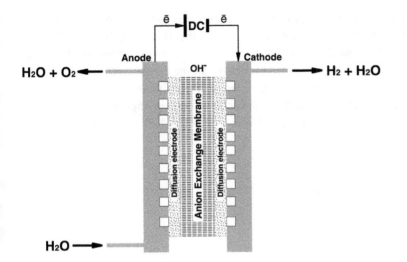

Figure 4.37: Principle of AEM electrolyser.

The evaluated catalysts, based on nickel or cobalt, do not use precious metals. The electrode reactions are those of alkaline electrolysis and operating temperatures between 50 and 70 °C. The materials for the membrane may be benzyltriethylamine, for example, used as the functional group.

The German company Enapter is developing this technology and presented a compact electrolyser: the 2.4 kW EL 4.0 module has an output 0.5 Nm³/h (1.0 kg/day) at 35 bar. The modules can be grouped in containers to provide 90 kW power (2022 data) for refuelling stations or used a single or combined units for micro-grid (solar PV, batteries or supercapacitors, fuel cell and AEM electrolyser) or residential application (the German HPS PICEA unit provides electricity by combining an AEM electrolyser and a fuel cell in one module).

The Singapore company Sungreen planned an AEM pilot project (1 kg/day module) in 2022.

AEM electrolysis is a promising technology as it combines the benefits of both technologies: no noble metal as catalyst like PEM and no liquid electrolyte (membrane) like alkaline electrolyser.

4.8.2.1 Electrolysers for power-to-gas
For use in the power-to-gas concept, electrolysers must meet a number of basic criteria:
- high capacity (several hundred to several thousand Nm³/h)
- a wide operating range (at least 0–100%)
- fast response time

Electrolysers must be able to rapidly absorb production peaks of excess renewable electricity. As in the case of gas-fired power plants, this results in the need for high peak power with the possibility of a relatively short operating time at maximum power. But this is one of the conditions to recover all the excess electricity. However, a balance has to be found between the expected surpluses (peak power, duration etc.) and the installed electrolyser capacity in order to avoid over-capacity, i.e., electrolysers staying idle when no surplus is available or when electricity price is too high.

5 Power-to-gas strategies

5.1 Hydrogen transportation

Once excess renewable electricity is used to produce hydrogen, it must be used directly or indirectly. Each of these options meets the needs of different sectors of the economy. After purification and according to the planned application, the hydrogen produced must be transported in gaseous or eventually liquid form.

5.1.1 Compression of hydrogen

Whether used locally or transported, the volume of hydrogen must be reduced by either compression or liquefaction. Depending on the technology used, the hydrogen produced can already be compressed at the outlet of the electrolyser at pressures up to several tens of bars.

For higher compression rates, four main technologies are used:
– piston compressor,
– diaphragm compressor,
– ionic liquids and
– thermal compressor.

What pressure to specify? It will depend on the intended use of hydrogen (Figure 5.1).

5.1.1.1 Piston compressor
One or more pistons compress the gas (Figure 5.2). Several stages can be connected to achieve high pressures. The choice of suitable materials avoids any lubricant that can contaminate the gas.

The pressures can reach 1,000 bar (some models of the German company Andreas Hofer, acquired by Neuman & Esser, can compress gas up to 4,500 bar).

5.1.1.2 Membrane compressor
In this type of compressor (Figures 5.3 and 5.4), a piston moves a membrane that compresses the gas. High-performance models have metallic membranes. Several stages can be used to achieve high pressures (up to 1,000 bar in 2017).

5.1.1.3 Ionic compressor
The German company Linde started in 2002 to develop an ionic compressor which was commercialised in 2009 for hydrogen.

https://doi.org/10.1515/9783110781892-006

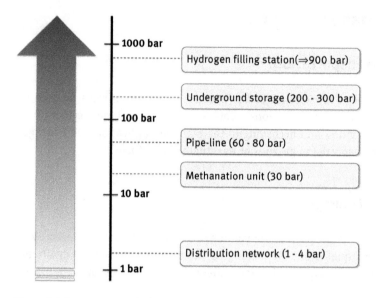

Figure 5.1: Pressure versus hydrogen use.

Figure 5.2: Piston compressor (Andreas Hofer Hochdrucktechnik GmbH).

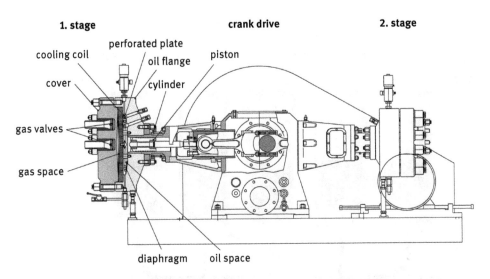

Figure 5.3: Membrane compressor principle (Andreas Hofer Hochdrucktechnik GmbH).

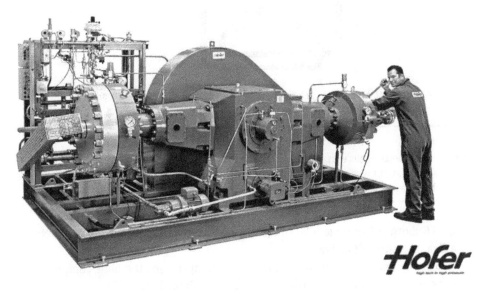

Figure 5.4: Single-stage membrane compressor (Andreas Hofer Hochdrucktechnik GmbH).

In this type of equipment (Figure 5.5), the gas is compressed by a practically incompressible ionic liquid replacing the pistons. A hydraulic pump moves the ionic liquid between two cylinders.

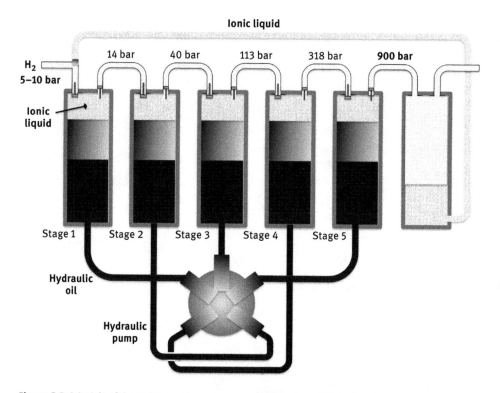

Figure 5.5: Principle of the multi-stage ionic compressor (according to Linde Gas documents).

Ionic liquids

Ionic liquids are organic salts with melting points below 10 °C. Unlike ordinary molecular liquids, they consist entirely of particles with negative and positive electric charges. They are not volatile or combustible and they have no measurable vapour pressure. They also do not mix with hydrogen.

Figure 5.6 shows a five-stage IC90 compressor with a maximum capacity of 370 Nm^3/h for a final pressure of 900 bar.

When compared with piston or diaphragm compressors, the ionic compressor is characterised by its small dimensions (1.1 m high for IC90; Figure 5.7).

One of the first units was installed at a hydrogen station in Switzerland existing in 2017 for fuel cell vehicles with 350 or 700 bar tanks. Other refuelling stations are using ionic compressors like the one in Venive/Italy or Pau/France.

5.1.1.4 Thermal compressor

Numerous studies have been done on the storage of hydrogen in hydrides. The next step was the possibility to reach high pressures at the hydride tank outlet.

Figure 5.6: Ionic compressor (Linde Gas).

The first compression is carried out during the filling under low pressure of the hydride tank at low or normal temperature. The desorption activated by heating increases the pressure of hydrogen at the outlet. Cyclic operation (low-temperature adsorption–desorption by heating) allows practically a continuous compression.

Developed by the Norwegian company Hystorsys with a first commercial model introduced in 2013, the hydride used absorbs hydrogen in monoatomic form in its crystalline structure and offers an important storage capacity.

The HYMEHC-10 (Figure 5.8) is a two-stage compressor system, which was able to supply continuously up to 10 Nm³/h hydrogen. In 2022, the HYMEC-12 had a capacity of 12 Nm³/h.

In the first stage, hydrogen is compressed from 10 to about 50 bar. In the second stage, the pressure is increased further to at least 200 bar. The alternating temperature cycle applies to both stages (low temperature of 20 °C and high temperature of 140 °C). The compressor is based on six metal hydride pressure vessels (three in each stage) and a controller running a time-based scheme for heating and cooling the vessels according to a defined sequence. Thus, by periodic heating/cooling (Figure 5.9) of the six vessels, continuous hydrogen compression is obtained.

Figure 5.7: Complete station with ionic compressor (Linde Gas).

The Greek company Cyrus and the French Eifhytec have also developed thermal compressor using hydrides. Both are still in the development/pilot phase with low output (1.5 Nm^3/h at 300 bar for the Cyrus compressor). This technology reduces considerably the number of mobile components outside the valves and is characterised by a silent operation. The systems are becoming more and more compact and the choice of hydrides makes it possible to adapt the equipment to the final use.

5.1.1.5 Balance of plant

The compressors described in this section require all the connections (gas, liquid, electrical) which make it possible to connect all the components to each other, to the electrolyser and to the following stations either for injection into the natural gas network or for an hydrogen service station (Figure 5.10).

Other auxiliary equipments are, for example, sensors or control and regulation electronics. Their operation also contributes to the optimisation of the electrolysis.

Figure 5.8: Two-stage thermal compressor (Hystorsys AS).

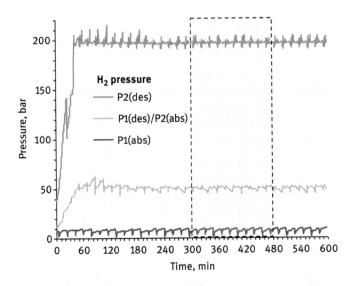

Figure 5.9: Compression cycles (Hystorsys AS).

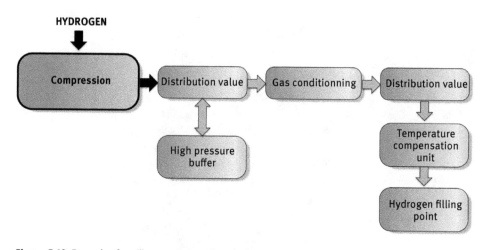

Figure 5.10: Example of ancillary equipment for a hydrogen service station.

5.1.1.6 Auxiliary equipment and performance

The auxiliary equipment for the optimal operation of an electrolyser can contribute to the efficiency of electrolysis if they are designed with maximum efficiency and reliability in order to allow an intensive use, especially during the peaks of excess electricity from renewable sources.

5.1.2 Hydrogen liquefaction

Industrial hydrogen liquefaction uses a variety of processes with helium, hydrogen or gas mixtures as coolant. The simplest method of liquefying hydrogen is that of Joule–Thompson (Linde cycle). For very large volumes, the cycles of Claude (pre-cooling with liquid nitrogen) and Brayton (helium or neon) are used nowadays.

The basic liquefaction method is based on the following steps: hydrogen is first compressed. It is then further cooled through multi-stage liquid nitrogen exchangers at 196 °C (78 K). Throttling through an expansion valve causes partial liquefaction (Joule–Thompson isenthalpic expansion). The remaining gaseous hydrogen is then returned to the compressor after passing through a heat exchanger where the cycle is repeated. The liquid hydrogen is stored in an insulated tank for further distribution.

***Ortho*- and *para*-hydrogen**

Hydrogen exists in two forms (*ortho* and *para*) according to the orientation of the spin of the nucleus (Figure 5.11). At room temperature, the *ortho/para* ratio is 75/25. In the liquid state, *ortho*-hydrogen slowly converts into *para* with a high energy release (527 kJ/kg), thus accelerating evaporation. To avoid this phenomenon, during liquefaction, *ortho*-hydrogen is converted into *para* by using a catalyst (iron oxide, activated carbon etc.).

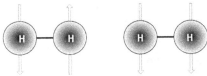

Para-Hydrogen   Ortho-Hydrogen   **Figure 5.11:** *ortho* and *para*-hydrogen.

Cryogenic hydrogen, referred to as liquid hydrogen (LH2), has a density of 70.8 kg/m^3 at normal boiling point (–253 °C), critical pressure being 13 bar and critical temperature –240 °C. LH2 has a much better energy density than compressed. The disadvantages are the evaporation losses (boil-off) and the need for super-insulated cryogenic containers.

Compression and liquefaction energies

Hydrogen compression at pressures between 200 and 700 bar requires an energy which depends on the final pressure and the chosen compression method (isotherm, adiabatic, multi-stage). This energy also depends on the initial pressure. For adiabatic compression (without heat exchange with the external environment), the compression energy is in the order of 10–16 MJ/kg of hydrogen for pressures of 200–700 bar, which represents between 8% and 12% of the energy content of hydrogen (higher calorific value). If the associated losses (consumption of auxiliary equipment) are taken into account the compression energy is 15–20 MJ/kg, i.e. 10–15% of the energy content.

The energy required for **liquefaction** of hydrogen represents a larger share of its energy content (LHV of 119.9 MJ/kg). If theoretically about 11.8 MJ/kg is sufficient to liquefy hydrogen (in the case of an ideal liquefaction process), depending on the size of the liquefaction unit, minimum values of the order of 40 MJ/kg are needed (for large capacity from 1,000 kg/h) or practically a minimum of 30% of the energy content.

5.2 Hydrogen transportation

5.2.1 Transportation dedicated to hydrogen

If the local use of hydrogen or its injection into the natural gas network is the most optimal solution, it can be transported in relatively large quantities in different ways:
- by pipeline,
- compressed in tanks or bottles and
- in liquid form.

The main networks of pure hydrogen pipelines are located in Europe (1,600 km) and in the USA (2,600 km) and are managed by the main producers (Air Liquide, France; Linde, Germany; and Air Products and Chemicals, USA). They represent a total of approximately 5,000 km in 2022. They link certain production sites directly to their users. In 2022, about 30 dedicated hydrogen pipeline projects are planned or projected by 2030/2040. The main use would be either for the industrial parks (steel, ammonia production etc.), hydrogen from wind or solar parks or for export (H2 Hub Gladstone in Australia or HHN in Chile, for example). In 2022, the French GRTgaz launched the RHYn project of a 100 km pipeline covering the upper Rhine and using to a large extent (up to 60 km) the existing natural gas network. It should be operational by 2028 and able to deliver 125,000 tonnes of hydrogen.

A network dedicated to hydrogen?

A study of the German research Centre Forschungszentrum Jülich of 2016 [1] quantified the construction of a hydrogen pipeline network for Germany. The proposed 40,000 km would be divided into a transportation network (12,000 km) and a distribution network (29,700 km). The cost of the project would be €18.7 billion with a transportation capacity of 2.93 million tonnes of hydrogen (about 33 billion Nm^3) per year to supply mainly a network of about 10,000 service stations the cost of which is not included in the project. A study by the National Renewable Energy Laboratory in 2012 estimated the cost between US\$ 2.8 and US\$ 5 million per service station. In 2022, the ambitious *European Hydrogen Backbone* is a project for hydrogen pipeline transport covering 28 countries [2]. The targeted 28,000 km pipeline in 2030 (53,000 km in 2040) should extend from South of Italy (with a North Africa link) to Norway. Part of it (40–60%) will use the existing natural gas network. Total cost in 2040 is estimated to be €80–143 billion. In 2021, the Port of Rotterdam started a feasibility study of a pipeline to Germany (*Delta Corridor*) especially if hydrogen will be imported to Rotterdam.

The transportation of compressed hydrogen is hampered by its low density even at high pressures. Road delivery per cylinder transports about 300 kg of hydrogen, i.e. 3,400 Nm^3 under 200 bar (Figure 5.12).

Figure 5.12: Transportation of gaseous hydrogen under 200 bar (Linde Gas).

The German company Wystrach has developed a range of containers for transporting hydrogen based on bottles. The basic module WyBundle (6–18 cylinders/200–1,000 bar) can be combined in 20–45 ft containers to provide large storage capacity for refuelling station, passenger trains etc.

Liquefied (which absorbs at least 30% of its energy content) hydrogen requires thermally insulated tanks to reduce losses by evaporation. The only advantage is the reduction in the cost of road transportation of hydrogen due to the higher density than that of compressed hydrogen. A 45 m^3 tank can deliver about 3.2 tonnes of LH2 (Figure 5.13).

Figure 5.13: Transportation of liquid hydrogen (Linde Gas).

The first world **vessel LH2 carrier**, the 8,000-gross ton diesel-powered Suiso Frontier (suiso meaning hydrogen in Japanese), from Kawasaki Heavy Industries Shipyard made its first liquid hydrogen delivery from Australia to Japan in 2022. It was part of the Hydrogen Energy Supply Chain and HySTRA program between Japan and Australia. Hydrogen will be produced from coal with carbon capture and storage (CCS) to reduce the environmental impact. The 9,000 km trip (23 days) allowed to deliver 1,250 m^3 of liquid hydrogen corresponding to about 2,950 MWh (LHV). Compared to LNG, the same volume would represent about 7,300 MWh (LHV). According to Kawasaki Technical review, the total cost of hydrogen is ¥29.3/Nm^3 (2.4 US$/kg), the liquefaction cost representing about 33%. At the city of Kobe, the hydrogen is also used to fuel a 1 MW gas turbine. Liquid hydrogen transportation over long distances is considered by many countries like Germany or Netherlands: several supply chain projects are under consideration or development.

5.2.2 Direct injection of hydrogen into the natural gas network

This is the most logical and technically simple option: purified hydrogen is injected into the existing natural gas network. This approach should ensure transparency for the end-user who should not alter the settings of natural gas equipment.

Hydrogen added to natural gas may affect the infrastructure through:
– hydrogen-induced stress or corrosion,
– increased safety risks during the transmission, distribution and use,
– degradation of performance of end-user equipment and
– degradation of performance or quality of industrial processes.

5.2.2.1 Natural gas network

As with the power grid, natural gas is transported to the end-user through a series of pipelines (Table 5.1). The diameters and pressures vary from 1.4 m up to 100 bar for the transportation network to 100 mbar for the residential, for example.

Table 5.1: Natural gas network.

Country	Total (km)	Transmission (km)	Distribution (km)
USA		490,000	2,000,000
Japan		5,000	250,000
China		110,000	
Germany	511,000		
France		32,000	200,000
UK		7,600	275,000

These networks allow the transportation and distribution of large quantities of natural gas, both annually (94 billion Nm^3 for Germany in 2021 or 130 for Japan) and in terms of flow (millions of Nm^3/h). In addition, they represent a non-negligible storage capacity. As an example, an 80 km section with a diameter of 105 cm under 70 bar contains 5.7 million Nm^3 of natural gas.

Hydrogen injection in the natural gas network is considered as the next step for use in industry, commercial or residential in order to reduce import of natural gas. Projects or roadmaps for 2030 or 2050 are established by numerous countries. Some experimentations have already been made. In France, the program GRHYD, started in 2014 and inaugurated in 2018, where hydrogen, supplied by a 50 kW PEM electrolyser, was injected during 21 months first 6% (10% then 20% in 2019, followed by a variable concentration with a maximum of 20%) in a local network. About 200 residential users and an hospital were fed with the hydrogen blend for cooking, heating and hot water. No equipment has been modified. In 2022, the feasibility has been proven (distribution network and equipments, lower emissions). In Germany, end of 2021, the

Avacon subsidiary of the energy supplier E.on has started a project in Saxony-Anhalt which covers 35 km of gas pipeline and 350 customers. Hydrogen injection will start with 15% mix, increasing to 20% in 2022/2023. Many national or international projects concerning hydrogen injection have or are being conducted: HyDeploy in the UK, European project ThyGA, HypSA in Australia, HyBlend in the USA etc. The European project Ready4H2, launched in 2022 and involving 66 DSO (Distribution System Operators), should prepare a roadmap for hydrogen injection in the European natural gas network. Very optimistic goals have been set: by 2040, 67% of network should be able to use 100% hydrogen feed.

The hydrogen injection in the natural gas network has however an important limitation: the volumes of green hydrogen needed. Hydrogen produced in 2022 is mainly from natural gas. Available green hydrogen volumes are still negligible compared to different set targets. In 2021, natural gas consumption for Germany was about 91 billion m^3. Injecting 20% of green hydrogen would require 18 billion m^3 of hydrogen. The total worldwide hydrogen production in 2021 (reforming, electrolysis etc.) was only 1.1 billion Nm^3!

However, there are still long-term challenges: overall cost of the adaptation to increased hydrogen concentration, behaviour of steel or plastic over long time etc.

Hydrogen diffusion and embrittlement in steel alloys

Given the small size of the hydrogen molecule, any mechanical or crystalline defects in natural gas pipeline, gas engines or gas turbine can result in the diffusion of hydrogen into the structure where it is in atomic (H^+) form. Hydrogen can also be present in the manufacturing process (casting of steel, welding operations etc.). This imprisoned hydrogen can lead to brittleness of steels. However, their surface undergoes a passivation which reduces the risks of diffusion. It can occur if defects are formed (extreme variation in temperature, high mechanical stress etc.). This risk increases with the concentration of hydrogen in natural gas.

5.2.2.2 Percentage of hydrogen injectable

The maximum level of hydrogen injection into the natural gas network is not specified for all countries. It is very variable without a valid scientific or technical reason being always advanced (Table 5.2).

Table 5.2: Limit of injection of hydrogen into the natural gas network in mol%.

Country	Belgium	UK	Sweden	Austria	Switzerland	Germany	France	Netherlands
	0	0.1	0.5	4	5	5–10	8	12

A necessary harmonisation, based on technical and scientific criteria, would allow injections and exchanges at European level, for example, as it is only possible on the basis of the country that has specified the lowest concentration of hydrogen.

5.2.2.2.1 A standard for vehicles using natural gas

In Europe, the hydrogen concentration for natural gas used as fuel (CNG – compressed natural gas) is limited to 2% in 2017 (German standard DIN 51624 or UN ECE R 110 specification for steel tanks).

5.2.2.3 Criteria for reliable injection

Numerous parameters have to be verified from the injection station to the final user (households, industry, energy etc.):

- transportation network (pipes made of metal or plastic): possible embrittlement in materials,
- distribution network: compatibility of compressors and measuring or control devices,
- transmission and distribution network: hydrogen loss assessment,
- households: compatibility of domestic equipment (e.g. boilers),
- energy: compatibility of gas turbines or cogeneration units and
- transport: compatibility with natural gas vehicles above 2%.

Separately, each parameter can allow the injection of high levels of hydrogen (current gas stoves tested with up to 30% hydrogen) and natural gas networks allowing up to 10%. However, a total compatibility for all users whatever the injected level must be guaranteed.

All necessary adaptations (sensors, ranges or boilers, transportation, distribution pipe equipment etc.) may also lead to significant costs of the order of several hundred million euros for Europe.

Separation of injected hydrogen from natural gas

Another option is, depending on the intended use, the extraction of injected hydrogen from natural gas. This option requires an additional step. The available technologies include:

- pressure swing adsorption (PSA),
- separation by membrane,
- cryogenic process,
- electrochemical separation and
- In addition to the costs of equipment and treatment, it will be necessary to reinject or use the separated natural gas.

The 2017 Austrian program *HylyPure* [3] studied the separation of hydrogen injected into the natural gas network. Austria allows a hydrogen concentration of up to 4%. The process consisted of three steps: pre-separation by selective polymer membrane, enrichment by PSA and then purification. The recovered hydrogen has a purity of 99.97%, compatible with fuel cells for hydrogen vehicles. In 2022, the German companies Linde Engineering and Evonik started a full-scale plant to separate hydrogen from natural gas using HISELECT® membrane. The hydrogen concentration can reach up to 60% and the purity is higher than 90%. A further step using PSA could increase the purity to 99.9999%.

5.3 Hydrogen storage

Faced to the very low gaseous density and also the very low liquefaction temperature, storing hydrogen presents numerous challenges: how to provide the highest energy content at the lowest cost, the easiest way of transportation, the lowest possible volume of the infrastructure. Different approaches are possible (Table 5.3).

Table 5.3: Hydrogen storage strategies.

Class	Technology
Physical	Compressed (tank)
	Compressed (caverns, natural gas reserves)
	Liquid
	Cryo-compression
	Slush (liquid + solid)
Chemical	Hydrides
	Adsorption (MOF, CNT)
	Liquid organic hydrogen carrier (LOHC)
	Hydrolysis (borohydrides)
	Reformed fuels, methanol, ammonia etc.

The chemical "storage" as reformed fuels (e-fuels) or chemicals like ammonia and methanol requires further processing steps whereas the other methods store hydrogen directly.

5.3.1 Compressed in tanks

This solution is adapted for transportation or relatively small volumes as it is the simplest, but given the low density of hydrogen, it would require high storage volumes even under high pressures. It would be incompatible for quantities produced with surplus electricity with the power-to-gas.

5.3.2 In liquid form

The energy required to liquefy hydrogen is too important compared to the energy content, resulting in a low overall efficiency. In addition, transportation and storage require specially designed and expensive tanks. Liquid hydrogen applications are also limited (space launchers, some filling stations for fuel cell vehicles, industry and research). This approach is tested for the transportation of large volumes over long distances (Australian hydrogen for Japan, for example).

Cryo-compression combines liquid hydrogen and high pressure cryogenic tank (typically less than 350 bar). It delivers the highest storage density. The boil-off is reduced: stored hydrogen can be kept many days without losses. However, the infrastructure costs (cryogenic and pressure resistant tank) are a major hurdle.

Slush hydrogen is a mixture of liquid and frozen hydrogen in equilibrium with the gas at the triple point, 13.8 K. The density is about 15–20% higher than that of the liquid and a longer storage life is possible.

Cryo-compression and slush hydrogen are considered for transportation due to the high storage density. The German company Cryomotive GmbH is developing cryo-compressed hydrogen tanks for trucks. Up to 115 kg of cryo-compressed hydrogen could be stored in four tanks behind the cabin. Target is 2025 for first deployment of storage systems and refuelling.

5.3.3 In metal hydrides

Some alloys can combine with hydrogen to form metal hydrides: MgH_2, Mg_2NiH_4, $LaNi_5H_4$, $NaAlH_4$ etc. The absorption of hydrogen takes place at the surface of the metal or alloy (physical and/or chemical adsorption) and then it diffuses into the structure of the crystal lattice (Figure 5.14). These hydrides can thus store hydrogen in their structure and release it if heated.

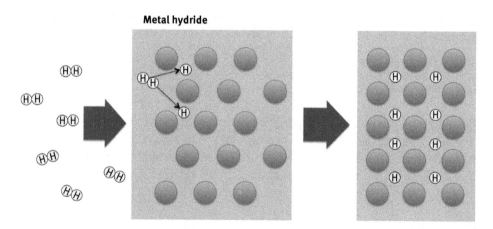

Figure 5.14: Principle of storage of hydrogen in hydrides.

The storage of hydrogen in hydrides does not require high pressures and delivers the stored hydrogen at constant adjustable pressure. The counterpart is the weight of the system: the stored hydrogen represents a few per cent of the mass depending on the hydride used; the theoretical values (Table 5.4) are rarely reached. The desorption temperatures of the compounds with useful storage ratios (>5%) are also relatively high.

Table 5.4: Properties of some hydrides.

Hydride	Hydrogen storage % mass	Desorption temperature at 1 bar in °C
La Ni$_5$H$_6$	1.4	25
TiV$_2$H$_4$	2.6	40
ZrMn$_2$H$_3$	1.8	167
NaAlH$_4$	5.0	220
MgH$_2$	7.6	>300

This results in an important weight of storage units using hydrides, such as those installed, in German submarines *Type U 212* for supplying fuel cells. Eighteen tanks weigh each 4.4 tonnes and contain 55 kg of hydrogen (620 Nm^3), representing 1.25% of the total weight. The new generation of this type of submarine is still being built.

When filling under pressure (generally 20–30 bar), the reaction being exothermal, it is necessary to cool the system. The stored hydrogen is released by increasing the temperature of the system as a function of the hydride (Figure 5.15). Thermal flux management is therefore the most critical parameter to manage.

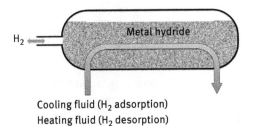

Cooling fluid (H$_2$ adsorption)
Heating fluid (H$_2$ desorption)

Figure 5.15: Hydrogen storage and recovery.

The compromise to be found lies between the storage capacity (weight of H$_2$/weight of hydride), the release conditions (temperature and pressure) and the stability over time of the hydride.

Among the hydrides used, the sodium alanate (NaAlH$_4$) makes it possible to illustrate the release mechanism under the effect of heat:

$$3\,NaAlH_4 \rightarrow Na_3AlH_6 + 2\,Al + 3\,H_2 \tag{5.1}$$

$$Na_3AlH_6 \rightarrow 3\,NaH + Al + \frac{3}{2}H_2 \tag{5.2}$$

The storage of hydrogen in hydrides, nevertheless, offers an increased safety with respect to gas or liquid storage.

The GKN HYDROGEN company, a subsidiary of GKN Powder Metallurgy, offers different storage containers (Figure 5.16): HY2MIDI (10 feet container up to 25 kg/ 420 kWh), HY2MEDI (20 feet container up to 120 kg/2 MWh) or HY2MEGA, the largest available unit in the market in 2022 (up to 250 kg/8.3 MWh). The basic storage module consists in pellets hydride assembled in a tank. Tanks can be stacked in order to achieve the storage volume required. Working pressure is less than 40 bar and desorption (hydrogen release) requires a temperature of 45–65 °C. The modules are used for off-grid applications, charging stations for EV, backup system etc.

Figure 5.16: A 250 kg/8.3 MWh hydride storage HY2MEGA, module with hydride pellets, 10 ft container HY2MINI (GKN HYDROGEN).

If hydrides offer advantages in terms of storage safety (e.g. no high pressures) or occupied volume compared to reservoirs (hydrogen gas or liquid), the main disadvantage is the weight of the system (reservoir and hydrides) (Figure 5.17).

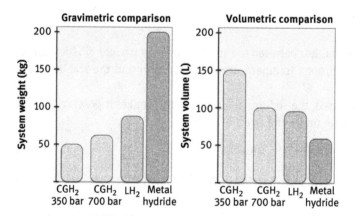

Figure 5.17: Comparison of storage capacities for 3 kg H_2.

5.3.4 Storage in caverns

This solution allows to store large quantities of hydrogen in the underground. In addition to existing facilities, studies and simulations have also been based on saline caves with volumes up to 500,000 m^3 (Figure 5.18). Hydrogen can be stored at pressures up to 200 bar.

Figure 5.18: Storage of hydrogen in a cave (KBB Underground Technologies GmbH).

According to the German company *DEEP.KBB GmbH*, such a cavern in a saline layer about 1,000 m deep would store 4,000 tonnes of hydrogen under a pressure of 10 bar. The corresponding energy stored would be 133 GWh. Storage efficiency would be about 98%. The costs of geological research, creation of the cavern and associated facilities would be in the order of €90 million in 2016. DEEP.KBB is part of the H2CAST

(H2 CAvern Storage Transition) project in Etzel, Germany that will explore starting 2022 the hydrogen storage in a cavern with injection in 2024.

An installation exists in England at Teeside with three caverns located at 370 m depth, each capable of storing 70,000 Nm^3 of hydrogen under a pressure of 45 bar. Austria launched an "Underground Sun Storage" programme in a cavern used to store natural gas. Hydrogen is supplied by four alkaline electrolysers (1.2 MW) with a production of up to 250 Nm^3/h which is then mixed with natural gas and injected. In Texas, Air Liquide and Praxair store hydrogen in caverns. Other facilities of relatively small capacity exist or have been used in Germany, Russia and Czechoslovakia and other countries.

Experimentations are taking place in the USA: *Advanced Clean Energy Storage*, in the State of Utah, 520,000 m^3 of storage capacity, combined with a 220 MW alkaline electrolyser and feeding starting 2025 a 840 MW combined cycle gas turbine (CCGT) or in Germany (H2 UGS, part of the HYPOS project, 7 million m^3 of capacity/126 GWh). Many other trials have started or are planned: European programme HyPSTER with storage in a French site; EWE in Germany, in the frame of the HyCAVmobil (Hydrogen Cavern for Mobility) with a test site near Berlin (550 m^3 in a first step); the HYBRIT programme in Sweden in a rock cavern with a pilot of 100 m^3.

This approach, if it allows to store large quantities of hydrogen for a recovery, could only be conceived in an area close to electrolysers in order to avoid building pipelines dedicated exclusively to hydrogen over long distances. Moreover, the geology of the underground must either has a structure allowing the creation of such caverns or provides adequate existing ones.

5.3.5 In natural gas reserves

5.3.5.1 Storage of natural gas

In order to benefit from a strategic reserve in the event of a breakdown in delivery or high demand, each country has large volumes of natural gas storage sites, generally and mostly in natural caverns. In 2020, over 600 facilities were in service worldwide with a total capacity of 420 billion m^3. The addition of hydrogen to natural gas would allow it to be stored without specific investments.

The available storage capacities for large volumes of hydrogen will depend on the percentage accepted (Table 5.5).

By assuming the hydrogen production energy at 5 kWh/Nm^3, each million normal cubic metre produced would store the equivalent of 5 GWh of surplus electricity. For a hydrogen injection rate of 5%, Germany could store 5.5 TWh of electricity and 33 TWh for the USA. Furthermore, this storage could be carried out continuously, the hydrogen–natural gas mixture being then injected into the network for use.

Table 5.5: Potential storage capacity of hydrogen mixed with natural gas (million m³).

	Natural gas storage capacity (million Nm³)	Hydrogen injected		
		2%	5%	10%
USA	163,000	3,260	8,150	16,300
Russia	75,000	1,500	3,750	7,500
Germany	23,000	460	1,150	2,300
China	14,000	280	700	1,400
France	14,000	280	700	1,400

5.3.6 Other storage methods

5.3.6.1 Liquid organic hydrogen carrier
In this technology, the energy carrier is a liquid where hydrogen is indirectly stored through chemical binding with a compound like dibenzyltoluene (Figure 5.19) with a "storage" capacity of 624 Nm³ (57 kg) H_2/m^3 LOHC.

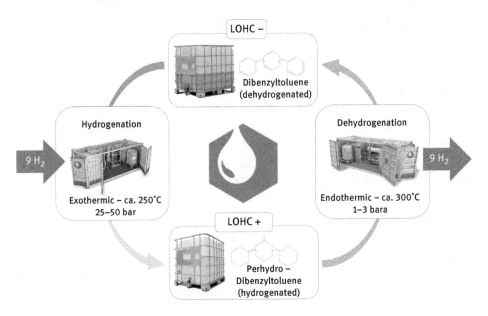

Figure 5.19: LOHC storage principle (Hydrogenious Technology).

When compared to storage under pressure, the advantages are reduced weight and volume and easier to handle: to store 3 kg of hydrogen under pressure (300 bar), bottles with a volume of 150 L and weighing 250 kg would be required, whereas the technology LOHC would store the same quantity in a volume of 50 L weighing 50 kg.

The German company *Hydrogenious Technology* has developed initially several storage (StorageBOX) or recovery (ReleaseBOX) units with different capacities (10–100 Nm^3/h storage). In 2022, Hydrogenious is proposing high volume storage plants (Figure 5.20) with a hydrogen capacity of 5–12 tonne/day (210–835 kg hydrogen/h) and a release plant with a capacity of 1.5 tonne/day (65 kg hydrogen/h). They are associated with storage tanks from 1 to 700 m^3 capacity. Smaller storage and release boxes with a capacity of 10 Nm^3/h are also available in containers. A large unit is being built in Germany (North Rhine–Westphalia) to store green hydrogen for industrial or mobility use. Planned capacity is 1,800 tonne/year. Another application with the Norwegian shipping company Johannes Østenjø dy AS is for propulsion of ships, associated to a fuel cell.

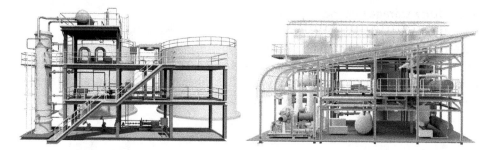

Figure 5.20: Storage and release plants (Hydrogenious Technology).

This approach allows hydrogen to be transported more easily and in larger volumes than in gaseous or liquid form and also with greater safety (e.g. supply of hydrogen service stations). Japan has started importing LH2 from Australia. The vessel Suiso Frontier can carry 1,250 m^3 of LH2 representing a weight of 88.56 tonnes that must be kept at −253 °C. Stored in LOHC, the same volume of hydrogen would take about 1,500 m^3 for a weight of about 1,500 tonnes but at room temperature.

5.3.6.2 Hydrozine™

The University of Eindhoven in Netherlands (*TeamFAST project*) has developed an alternative based on the use of surplus electricity for the conversion of water and CO$_2$, mainly into formic acid (HCOOH) through a new catalyst, supplemented by additives forming hydrozine™ with an energy density of 2 kWh/L:

$$2\bar{e} + CO_2 + H_2O \Rightarrow HCOOH + {}^1/_2 O_2 \tag{5.3}$$

The liquid can be stored and handled at room temperature for industrial or transportation applications. In the latter case, the formic acid makes it possible, after reforming, to recover hydrogen to supply a fuel cell. With the Dutch transportation company VDL, a demonstrator for a bus equipped with range extender (25 kW fuel cell) in a trailer was presented in 2017 (Figure 5.21).

Figure 5.21: Range extender using Hydrozine (TeamFAST).

The TeamFAST moved in 2018 to the creation of a company, DENS. After a 2 kW version of the X2 generator, the commercial unit is located in a 10 feet container. The 4,000 L tank can store up to 3.2 MWh and generate 25 kW nominal power through a fuel cell. Such unit can be used as zero-emissions generator for events, construction sites, emergency power etc.

5.3.6.3 PowerPaste
This compound is based on magnesium hydride which reacts with water (exothermic reaction) by releasing 1,700 L of hydrogen per kilogram of hydride. The challenge is to find an equilibrium for the hydrolysis reaction to provide a hydrogen flow rate function of demand. The use of additives makes it possible to form a paste that is easy to handle. The German institute *Fraunhofer IFAM* [4] has developed two demonstrators around this technology to supply a fuel cell (50 or 300 W). However, this system is linked to a distribution network of PowerPaste which is consumable and remains also the management of the residues, mainly of magnesium hydroxide.

5.3.6.4 Other methods
Studies have been or are being carried out by research institutes on the possibility of industrial storage of hydrogen in clathrates, metal organic framework, nanomaterials or graphene. None of these technologies has reached the stage of demonstrator.

5.4 Conversion of hydrogen into methane: methanation

5.4.1 Thermochemical methanation

Methanation (Figure 5.22) is the process of converting hydrogen into methane according to the reactions:

$$H_2 + CO_2 \Leftrightarrow CO + H_2O \tag{5.4}$$

$$CO + 3H_2 \Leftrightarrow CH_4 + H_2O \tag{5.5}$$

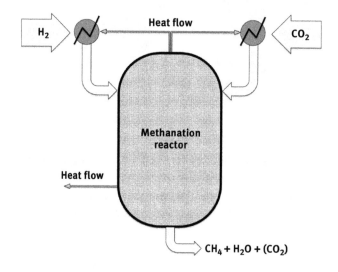

Figure 5.22: Thermochemical methanation.

The overall reaction is

$$CO_2 + 4H_2 \Leftrightarrow CH_4 + 2H_2O. \tag{5.6}$$

The reaction is exothermic (heat release: $\Delta H = -164.9$ kJ/mol under normal conditions) and takes place in the presence of a catalyst generally in the form of pellets through which the gases pass.

Type of thermochemical methanation reactor
- in fixed bed methanation, the reactor is packed with the catalyst with a particle size in the range of millimetres
- in fluidised bed methanation, fine catalyst particles are fluidised by the gaseous reactants
- in three-phase methanation, a solid catalyst (powder <100 μm) is suspended in a stable temperature inert liquid as dibenzyltoluene

The yield is of the order of 80% with a methane content greater than 90%. This conversion rate depends on temperature and pressure. This reaction has been known since 1905: the chemist Paul Sabatier experimented it with the use of nickel catalyst.

Methanation can be carried out without direct temperature control (which can rise to 700 °C) using a series of reactors with gas cooling between each. The other option is to regulate the temperature of the reaction by using a cooling system and recovering the heat released (Figure 5.23).

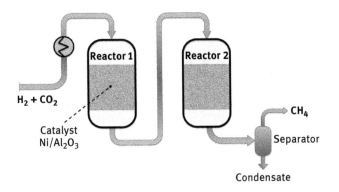

Figure 5.23: Lurgi methanation steps (simplified).

At the reactor outlet, a mixture of methane and hydrogen is obtained. The methane obtained (synthetic natural gas) can be injected into the natural gas network after treatment in order to have a hydrogen level corresponding to that allowed.

5.4.1.1 Overall yield of thermochemical methanation
Like any transformation, the yield of this step (Figure 5.24) affects the final one of the power-to-gas technology.

The best conversion rates for methanation alone reach 80%. A recovery of the released heat makes it possible to improve the total efficiency.

5.4.2 Co-electrolysis

It is the combination of high-temperature electrolysis (SOEC) with CO_2 injection, followed by methanation using the CO/H_2 mixture supplied (Figure 5.25).

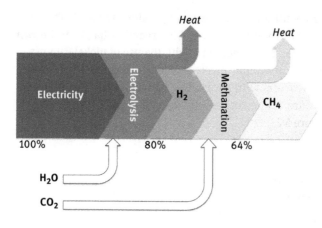

Figure 5.24: Performance of the power-to-gas chain with methanation.

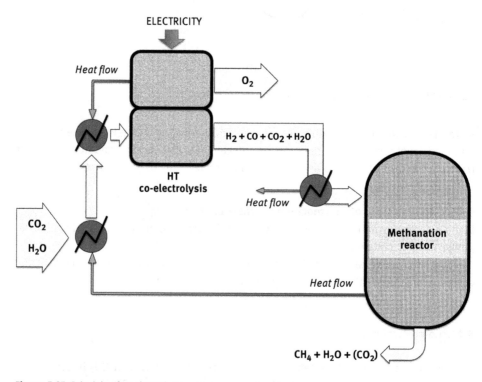

Figure 5.25: Principle of co-electrolysis.

Two reactions occur in the high-temperature electrolyser (Figure 5.26):

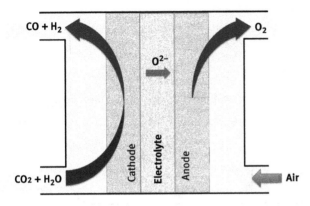

Figure 5.26: Co-electrolysis cell.

Decomposition of water:

$$2\,H_2O \Leftrightarrow 2\,H_2 + O_2. \tag{5.7}$$

Reverse shift reaction:

$$CO_2 + H_2 \Leftrightarrow CO + H_2O. \tag{5.8}$$

Methanation is carried out in a reactor in the presence of nickel:

$$CO + 3\,H_2 \Leftrightarrow CH_4 + H_2O. \tag{5.9}$$

The overall reaction is

$$CO_2 + 2\,H_2O \Leftrightarrow CH_4 + 2\,O_2. \tag{5.10}$$

The development of co-electrolysis is based on chemical equilibrium models to evaluate the composition of the mixture produced as a function of the proportions of CO_2 and water injected as well as the parameters of the electrolysis (operating temperature, current density, voltage etc.).

High-temperature electrolysis and methanation coupling offer a better yield but some limitations (cost, degradation, development of new materials etc.) remain to be overcome.

5.4.3 Biological methanation

In this approach, selected microorganisms use hydrogen and carbon dioxide to produce methane and water (Figure 5.27). This reaction takes place under atmospheric pressure and low temperature (40–70 °C) compared to the conditions of thermochemical methanation.

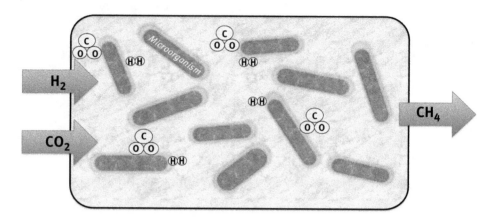

Figure 5.27: Principle of biological methanation.

The *Electrochaea* company, based on developments at the University of Chicago, was founded in 2010. In Denmark, it has set up in 2014 a biological methanation unit within the framework of the BioCat project (Figure 5.28).

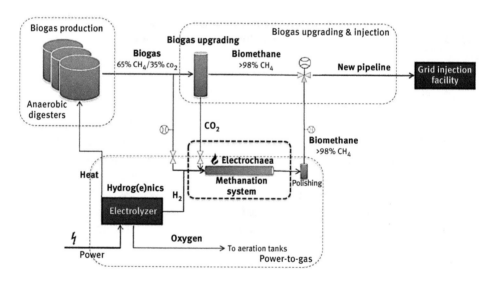

Figure 5.28: Biological methanation – BioCat project (BioCat).

The microorganism used (Archaea, Figure 5.29) is monocellular without genetic modification. The reactor temperature is 60–65 °C. In this project, conversion yield to methane is greater than 98%. The first phase in 2012 on the location of a biogas unit at Folum saw the implementation of a 4,800 L bioreactor which operated for 3,600 h with methane production ranging from 0.5 to 18 Nm^3/h.

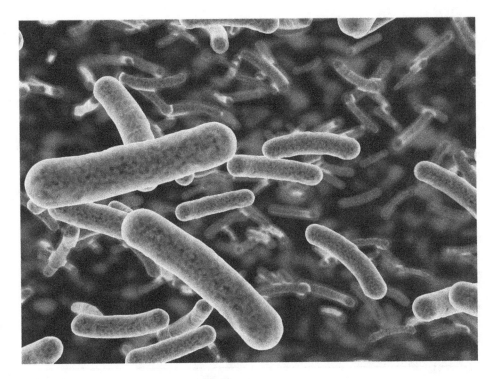

Figure 5.29: Archaea microorganisms (Electrochaea).

An industrial unit in Avedøre, with a production capacity of 50 Nm^3/h, integrated into a biogas unit, with a 1 MW electrolyser benefiting from an optimised reactor was tested in 2015. The first methane production took place in April 2016, 4 h after introduction of the microorganisms. The final mixture contained 97% methane. Other pilot plants were installed in Solothurn, Switzerland and in Golden, Colorado. In Belgium, the CO_2 from lime industry and hydrogen from a 75 MW electrolyser will be used to produce methane for injection in the natural gas network. Production start is planned for 2025. Biocat methanation plants can have a power methanation between 45 and 1,600 kW (0.9–44 m^3 of methane produced) requiring, respectively, 3.5–180 m^3 of hydrogen and an associated electrolyser of 1 to 50 MW.

MicrobEnergy, founded by the German company Viessman, has been bought in 2021 by the Swiss-Japanese Hitachi Zosen Inova. MicrobEnergy founded in 2012, has evaluated since March 2015 a demonstration unit (see chapter 7.3.1.3 for a detailed description) in Allendorf in Germany near a sewage treatment plant supplying the CO_2. The proton exchange membrane (PEM) electrolyser installed (300 kW with two stacks of 150 kW) allows the production of 15 Nm^3/h of methane with a purity of 98%, the remainder being mainly hydrogen. In 2022, an industrial scale unit using the BiON process has been commissioned in Dietikon, Switzerland. Hydrogen from a 2.5 MW electrolyser (450 Nm^3/h) allows the production of an equivalent of 18 GWh of methane, injected in the natural gas network.

Compared to thermochemical methanation, biological methanation allows, among others:

- integration into biogas plants (several thousands in Germany) to use CO_2,
- better catalyst management (biological self-reproducing),
- excellent yield (75–80%) and
- operation at lower temperature.

Origin of CO_2

The CO_2 needed for any methanation or co-electrolysis reaction can come from different sources:
- Recovery after purification of biogas
- CO_2 capture of industrial or power generation emissions
- Air sampling despite a low concentration (about 400 ppm)

The Swiss company *Climeworks* has developed an industrial process for recovering CO_2 from air.

Through a cyclic adsorption/desorption process, the atmospheric CO_2 is absorbed into a material while all other air molecules pass through (CO_2 free air is released). Once the filter is saturated, CO_2 is released at 95 °C. After CO_2 recovery, the absorbing material can be reused for other cycles. Commercial units are scalable modules with a daily capacity of 135 kg CO_2/day each. In 2017, an installation using 18 modules produces about 900 tonne/year CO_2 for a greenhouse. In Iceland, 4,000 tonnes of CO_2 per year are captured and pumped in the underground after mixing with water. Another project (Mammoth) should allow a capture capacity of 36,000 tonne/year.

5.4.4 Methanisation and synergy with power-to-gas

Methanisation consists in using organic waste, sewage sludge, for example, to produce, by *anaerobic fermentation* of microorganisms, a gas mixture whose most important constituents are methane (50–85%) and carbon dioxide (15–50%). After purification, methane is injected into the natural gas system or is used directly to supply a natural gas vehicle service station or used locally for a cogeneration or CHP unit with recovery of the heat produced.

In the power-to-gas approach, CO_2 produced by the methanisation unit can be used in methanation units (Figure 5.30).

5.4.5 Methanation, the key to electricity from renewable sources?

This technology allows the conversion of hydrogen produced by electrolysis using excess electricity into methane. It can then be injected in the natural gas distribution network or storage units. Even if the efficiency of power-to-gas is reduced through this approach, the integration of methane in the energy circuit does not require any modification of the existing procedures and installations.

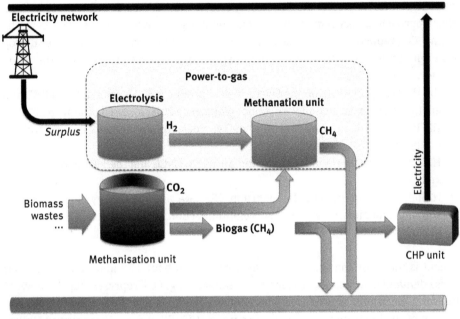

Figure 5.30: A methanation unit coupled to a methanisation unit.

5.5 Use of hydrogen or methane produced

Hydrogen and methane are commodities already used by different sectors of the economy. The additional production due to power-to-gas may open up other prospects or have an influence on the markets (price of these gases, for example, or development of new uses).

The production of hydrogen from surplus renewable electricity must make it possible to replace partly or completely the hydrogen produced by hydrocarbons reforming and/or eventually open up other utilisations. For each TWh of surplus electricity, it is possible to produce 200 million Nm^3 of hydrogen (hypothesis of 5 kWh/Nm^3).

5.5.1 Industry

The most important user of hydrogen is the petrochemical industry mainly for desulphurisation. Another important use is the production of ammonia for fertilisers.

Other sectors like steel, cement or glass industry are experimenting the use of hydrogen to reduce CO_2 emissions. Iron or steel industry accounts for about 7% of world CO_2 emissions in 2020 but almost 30% of industrial emissions. Using blast furnace/

basic oxygen furnace steel production, hydrogen can be used as reducing agent. Another approach, direct reduction of iron, would also allow the use of hydrogen. The optimal CO_2 reduction can be reached only when using green hydrogen. Several projects or pilot units, sometimes with integrated electrolyser, are planned for production between 2025 and 2030:

- the HYBRIT project involving Swedish steel producer SSAB, iron maker LKAB and energy supplier Vattenfall with a pilot plant since 2018, a demonstrator in 2026;
- H2FUTURE with the Austrian steel maker Voestalpine, Siemens Energy and the utility VERBUND;
- H2ERMES in Netherlands with the Indian TATA, HyCC and the Port of Amsterdam; and
- H2Stahl in Germany with Tyssenkrupp, NUCERA and the energy supplier STEAG with a production in 2025 after successful injection of hydrogen in a blast furnace in 2019.

Cement is the most consumed material worldwide (about 4.2 billion tonnes in 2021) whose demand is increasing. During its manufacturing, CO_2, representing about 6% of the world emissions, is produced when heating the kilns and during thermal decomposition of the limestone. The Mexican company CEMEX and the UK company HiiROC (hydrogen production by thermal plasma electrolysis), after an hydrogen injection trial in Spain in 2019, is feeding all European plants with an hydrogen mix. In 2021, Hanson UK, a subsidiary of Heidelberg Cement, operated a kiln with 100% hydrogen.

Although emitting less CO_2 than steel or cement industry, the glass manufacturing also has potential to reduce drastically CO_2 emissions. The Japanese Nippon Sheet Glass made in 2021 a trial using hydrogen at its UK facility in the frame of the HyNet Industrial Fuel Switching project. The German company Schott is planning to produce by 2030 environmentally neutral glass using hydrogen and launched in 2022 a project to gradually test a blend natural gas/hydrogen.

For the industry, switching to hydrogen requires high initial investments but OPEX will be determined mainly by the cost of green hydrogen compared to current fuels. Hydrogen, excluding petrochemicals and ammonia production, serves as the basis for many sectors like fine chemicals and electronics. Even if the quantities are relatively small, the purity of hydrogen produced by electrolysis and the possibility of producing it locally or even on site would be an advantage.

5.5.2 Energy: conversion into electricity

This approach, which could be called **P2G2P** (power-to-gas-to-power), involves generating electricity from hydrogen or methane produced through electrolysis or methanation. For this conversion, the existing technologies cover a whole range of power: CHP units, fuel cell or gas-fired power plants.

5.5.2.1 Fuel cells

This technology works on the same principle as PEM electrolysis with the same structure, but hydrogen and oxygen (or air) injected react to produce electricity (Figure 5.31). The principle was discovered in 1838 by the Swiss Schönbein and by the English Grove in 1839 [5]. It was only in the mid-1960s that this technology has been developed by NASA for the first space flights and towards the 1990s that high-power fuel cells reaching the power of megawatts have become available.

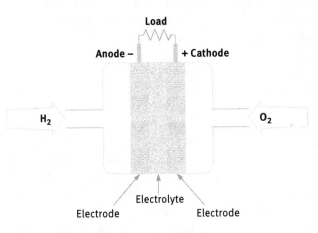

Figure 5.31: Principle of the fuel cell.

5.5.2.2 Types of fuel cells

Fuel cells are distinguished mainly by the type of electrolyte and the operating temperature (Table 5.6).

Table 5.6: Specifications of fuel cell types.

Type	Average temperature (°C)	Fuel anode	Ions	Fuel cathode
PEMFC	80–120	H_2	$H^+\Rightarrow$	O_2 or air
DMFC	110	Methanol	$H^+\Rightarrow$	O_2 or air
PAFC	200	H_2	$H^+\Rightarrow$	O_2 or air
AFC	80	H_2	$\Rightarrow OH^-$	O_2 or air
MCFC	650	H_2	$\Rightarrow (CO_3)^{2-}$	O_2 or air
SOFC	800–900	CH_4	$\Rightarrow O^{2-}$	O_2 or air

Their denomination reflects the electrolyte used:
- PEMFC (proton exchange membrane fuel cell)
- DMFC (direct methanol fuel cell)
- PAFC (phosphoric acid fuel cell)
- AFC (alkaline fuel cell)

- MCFC (molten carbonate fuel cell)
- SOFC (solid oxide fuel cell)

Fuel cells operating at low temperatures (<200 °C) use precious metals (platinum) as catalysts at both electrodes. The same problem as for the PEM-type electrolysers arises concerning the cost of these catalysts and the actual impossibility of replacing them. High-temperature fuel cells (MCFC or SOFC) do not require them, but use special alloys.

Even if some types can use hydrogen (e.g. SOFC), they use generally natural gas or methane. Others such as PEMFC and AFC can use methane, but by feeding it first in a reformer for conversion into high purity hydrogen that can then be injected in by the cell.

High-power fuel cells (more than 100 kW) include PEMFC (Ballard, Canada; Cummins/Hydrogenics, Canada; PowerCellution, Sweden; or Nedstack, Netherlands), PACF (Doosan Fuel Cell America, USA), MCFC (FuelCell Energy, USA), AFC (AFC Energy, UK) and SOFC (Bloom Energy, USA; Mitsubishi, Japan).

The Ballard fuel cell ClearGen II (PEMFC, 1.5 MW) can also be installed on a truck trailer The Nedstack PemGen MT-FCPI-500 with a nominal power of 500 kW_e and 400 kW_{thl}, corresponding to an hydrogen consumption of 59 MWh/MW_e, fits in a 20 feet container (total weight of 15 tonnes). The Nedstack PemGen CHP-FCPS-100 has a nominal power of 1,000 kW_e and 800 kW_{th}. It is built in a 40 feet container. The basic MCFC fuel-cell module from Fuel Cell Energy has a power of 1.4 MW_e with a 47% efficiency. A Bloom Energy module (SOFC, 300 kW) has an efficiency of 52% with a hydrogen consumption of 17.3 kg/h.

The US energy supplier Daroga Power will install up to 33 MW of SOFC from Bloom Energy to optimise base load power (Figure 5.32).

5.5.2.3 Fuel cell efficiency

Depending on the type and the power, the electrical efficiency can reach 50–60% and more than 90% if the heat is recovered, the reactions in the fuel cell being exothermic.

5.5.2.4 Advantage of fuel cells

Compared to a CHP unit or a gas-fired power plant, a fuel cell has virtually no moving parts (apart from auxiliary equipment such as pumps, fans etc.). Units operating at low temperatures (<100 °C) have a very short start-up time.

In 2008, South Korean government launched a major programme of high-power fuel cell to stabilise the electricity grid while recovering heat. Several parks contain dozens of fuel cells (Hwaseong city had 21 fuel cells of MCFC type with 2.8 MW or a total of 59 MW and Busan had 70 PAFCs with 400 kW each). In 2022, the largest world fuel cell power plant was inaugurated at Incheon. It has a power of 79 MW and an annual capacity of 700 GWh. Doosan Fuel Cell supplied 149 fuel cell units. Other fuel cell plants are planned in South Korea.

Figure 5.32: Bloom Energy SOFC units to stabilise power (Daroga Power) and Nedstack 2 MW PEMFC.

5.5.2.5 Power plants or CHP units

Natural gas power plants or high-power CHP units can take advantage and use synthetic methane to operate.

For these two electricity generation technologies, the use of methane from methanation is more appropriate as it does not require modification of certain components or new settings for combustion.

In many countries, gas-fired power plants are an important element in stabilising electricity grids during peak demand periods. However, they are used less and less because of the decline in the price of coal (important shell gas production in the USA), which favours those using it. The possibility of having large quantities of methane could revive their use as they have a very short start-up time. Turbines for power plants are able to use hydrogen either as a blend with natural gas/methane or at 100% concentration:

- Siemens Energy's gas turbines have received the "H2-Ready" certification.
- The Italian company Ansaldo has developed a family of gas turbines (GT36-S6/369 MW) able to run with up to a hydrogen concentration of 70%.
- The Japanese company Mitsubishi has established a demonstration facility in Hyogo Prefecture to test hydrogen use for gas turbine (in 2025 turbines are planned to use 30% hydrogen).
- In 2020, a Kawasaki Heavy Industries turbine of 1,900 kW_e was successfully tested with 100% hydrogen. The modified combustor allowed to produce less NO_x (<75 ppm) than conventional turbines using hydrogen. The German energy supplier RWE is planning to install a 34 MW Kawasaki hydrogen turbine in Lingen, Germany to be operational in 2024. The Lingen site will produce hydrogen starting 2024 (100 MW of electrolysers expanded to 300 MW in 2026).
- The European industrial scale project HYFLEXPOWER launched in 2020 in France is aimed to demonstrate the feasibility of the power-to-gas-to-power concept. The gas turbine (Siemens Energy SGT-400) will use green hydrogen (up to 100% in 2023) produced by an electrolyser fed by excess electricity,
- For GE Gas Power, USA, the 7HA.02 turbine, developed in 2017, can use up to 20% hydrogen without modification. Ongoing project in 2022 should check the feasibility to increase the hydrogen concentration up to 100%.

5.5.2.6 Conversion efficiency

The electrical efficiency of the P2G2P approach varies according to the chosen pathway for hydrogen production and use (Figure 5.33). For a fuel cell, a CHP unit or a gas-fired power plant, the conversion efficiencies are at best 40% for stationary applications, but over 60% for CCGT. However, depending on the production or supply strategy decided, this approach may have an economic and/or strategic justification (e.g. security or stability of supply), be it at the level of energy producer or industrial users.

The overall efficiency can be significantly improved by recovering the heat produced. This recovery will be all the easier if the gas-to-power unit is installed near the users (local district heating).

5.5.3 Mobility

5.5.3.1 The hydrogen for road transport : still a utopia?

Hydrogen can be used for fuel cell vehicles (Figure 5.34). The current number is, however, still so limited that the amount of hydrogen needed remains and should remain small despite the development forecasts (which have always been very optimistic).

Although many manufacturers have embarked on this development in recent years, most of them have still pilot or small manufacturing units. By the end of 2021,

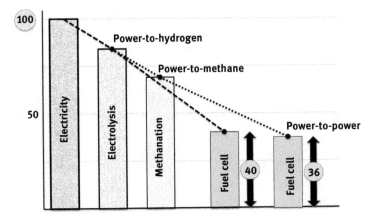

Figure 5.33: Comparative electrical efficiency.

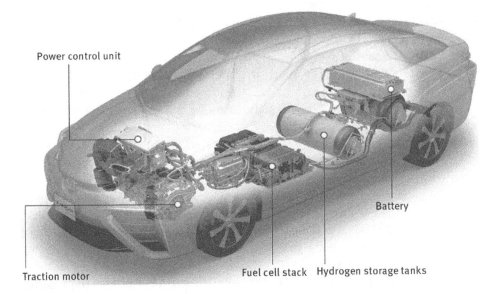

Figure 5.34: Mirai hydrogen vehicle and fuel cell (Toyota Motors).

35,000 fuel cell vehicles (all categories) were in operation worldwide (23,000 in Asia: Japan, South Korea and China) with 540 refuelling stations.

Passenger cars represented 75% of fuel cell vehicles in 2021. Most were part of a captive fleet (taxis or company vehicles, for example). The fleet of electric vehicles with batteries represented over 11 million vehicles in 2021. The number of important manufacturers of fuel cell passenger vehicles is still limited, mainly in South Korea and Japan (Hyundai, Toyota and Honda). The production is also very limited (5,700 Toyota Mirai and 8,900 Hyundai Nexo delivered in 2021) and costs are still high (about

60,000 US$ for the Toyota Mirai in 2022). Market is pushed by subsidies and exemptions. Another issue is the lack of sufficient refuelling stations especially for those not being captive. Even with those issues, the projections are still optimistic: in 2021, the International Energy Agency expects in 2030 about 10–15 million fuel cell cars and 500,000 trucks.

Hydrogen fuelled **busses** are still considered by many cities worldwide. Following the numerous programmes in the 2000s supported by subsidies, the development is still moving slowly. In 2022, the French city of Montpellier cancelled the order for 51 hydrogen busses for cost reasons. The total cost (including refuelling station) would have been €29 million with yearly operational costs amounting to €3 million versus €500,000 for battery busses. This is characteristic for the hydrogen busses market: high investment and operating costs, necessity to have an hydrogen infrastructure, manufacturers faced to issues like moving to large-scale production. Thus, only mainly through subsidies can actually the hydrogen busses be deployed. Whereas the number of battery busses is increasing: 8,500 were registered in 2021 in Europe and 385,000 in China. As a comparison, in 2022, there were less than 150 hydrogen busses operating in Europe and about 70 in the USA. The only country having a representative number is China with over 5,500 units in 2022.

The **truck** sector is faced to the same dilemma: batteries or hydrogen? In 2022, the number of light, medium or heavy duty hydrogen trucks was less than 3,500, almost all being medium duty. Truck manufacturers are still following the dual track or an hybrid version (PEM fuel cell + batteries). CellCentric (Daimler Truck-Volvo Group), Scania (Cummins fuel cell), Kenworth/USA, Toyota-Hino/Japan, Hyundai/South Korea and Nikola Corp. (Bosch fuel cell)/USA have shown prototypes, are developing or have already hydrogen trucks on the market. The main limitation is the high purchasing and OPEX costs over diesel trucks (more than twice in 2022). In Switzerland, Hyundai Motor has delivered 47 units in July 2022 and will supply a total of 1,600 XCIENT trucks by 2025. In July 2022, some of those trucks reached 180,000 km of mileage. The XCIENT truck (19 tonnes as rigid truck; Figure 5.35) has a range of 400 km. It is equipped with two 95 kW fuel cells and batteries (73 kWh). Hydrogen is stored at 350 bar in seven tanks with a capacity of 32 kg. On the other side, truck manufacturers are preparing series production of battery-powered trucks for 2024 while reaching cost parity with diesel trucks (falling costs of batteries). Those trucks will have a range of about 400 km and charging time will be possible during driver's mandatory breaks (45 mn every 4.5 h).

The other concern for fuel cell vehicles is the low overall efficiency of the hydrogen fuel cell pathway (Figure 5.36).

Figure 5.35: Hyundai XCIENT hydrogen truck (Hyundai Motor Company).

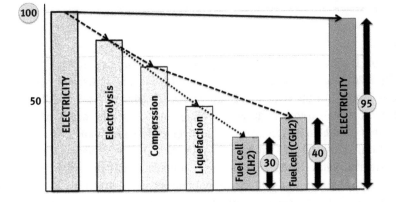

Figure 5.36: Yield of hydrogen/fuel cell vehicle (from well to wheel).

Range extender

Some companies offer an extension of autonomy for electric vehicle. These are compact fuel cells using hydrogen and providing extra electricity to increase driving range. If this solution seems, a priori, elegant, it does not solve the dilemma battery/hydrogen: high-speed charging is still too long and very few hydrogen refuelling stations are still available.

Experiments have been conducted with conventional vehicles (internal combustion engines) fuelled by hydrogen or a mixture of hydrogen–methane (hythane). In hythane, hydrogen generally represents 20% by volume. Compared to methane alone, the mixture reduces emissions (THC, CO and NO_x) as well as providing a better yield

(10–15%). The French programme ALT-HY-TUDE (2005–2008) allowed experimenting this mixture on buses. In India, the first filling station was opened in 2009 and in 2011 another at the San Francisco airport with a fleet of 30 vehicles. However very few projects followed.

Hydrogen for trucks or busses: total cost of ownership

The German VDI/VDE [6] and the French ADEME (Agence de l'environnement et de la maîtrise de l'énergie) [7] have published a comparison between different fuels/engines in 2022: fossil (diesel) versus battery versus hydrogen internal combustion engine (ICE)/fuel cell for the ADEME's report and battery (or catenary) versus hydrogen versus ICE (e-fuel) versus CNG/LNG for the VDI's/VDE's report. The French ADEME concludes that the best total cost of ownership (TCO) for a 44 tonnes truck or an 18 m bus over 12 years is reached by the diesel (considered as reference) whereas hydrogen has a TCO 1.5–3 times higher. For the VDI/VDE, the best efficiency is reached by the battery/catenary combination (up to 70%), the worst being ICE with e-fuel (up to 20%), although hydrogen/fuel cell being still in the low side (up to 29%). Batteries have an efficiency up to 63%. Both studies are based on 2021 prices (hydrogen, batteries etc.).

In the frame of the decarbonation of the transport, this shows the advantage of batteries for efficiency (catenary requires new investments) and their potential use for close distance delivery whereas hydrogen could be used for long distance transportation, providing that the price of fuel cell trucks or busses decreases.

5.5.3.2 Railway

In the railway sector, if a hydrogen train prototype was used in a mine in the USA in 2002, it was not until 2017 that a train was certified on a regular line in Germany (suppliers were Alstom, France, for the structure, Hydrogenics, Canada, for the fuel cell and Wystrach, Germany, for the gaseous hydrogen storage). Following this recent experimentation, the Alstom *Coradia iLint* (Figure 5.37) has been ordered by many Länder in Germany (over 40 units), by Italy (14 units) or France (12 units) or experimented by other countries (Austria, Netherlands, Sweden etc.). Siemens Mobility/Germany with the *Mireo Plus H* (1.7 MW power and 800 km range) and Stadler/Switzerland *Flirt H2* have also developed fuel cell passenger trains that will be on the market starting 2023–2024. Fuel cell trains have a range of 600–1,000 km and can be used on not electrified tracks, without replacing diesel trains thus reducing CO_2 emissions.

Like for trucks, the competition comes from battery-powered trains that can also run on electrified tracks and provide a lower cost alternative. The Stadler *FLIRT AKKU* has a battery only in the range of 200 km and the Siemens *Mireo Plus B* 120 km.

5.5.3.3 Maritime

Decarbonisation of the maritime sector is ongoing as it represents about 3% of CO_2 global emissions (13.5% in Europe in 2018). The International Maritime Organization (IMO) has set the goal to a 50–70% reduction by 2050 compared to 2008 levels. The technologies needed to reach this goal will depend on the type of vessel (ferry, fishing boat, cruise ships, carriers, oil tankers etc.) and energy supply (LNG, hydrogen, batteries etc.).

Figure 5.37: Hydrogen train Alstom Coradia iLint (top), battery trains Stadler *Flirt AKKU* (centre) and Siemens *Mireo Plus B* (Alstom, Stadler, Siemens).

EEXI and CII

The Energy Efficiency Existing Ship Index (EEXI) and the carbon intensity indicator (CII) are indicators from the IMO to reduce the greenhouse gases emissions from ships. The EEXI is calculated according to the design of the ship and is expressed in grams per tonne mile. The CII is a measure of the efficiency of a ship's cargo (goods, vehicles, passengers etc.) and is expressed in grams of CO_2 emitted per carrying capacity, and nautical mile and ships will be rated from A to E, A being the best. The use of LNG would improve the EEXI by 25%. The ships must comply with the defined EEXI and CII by 2023/2025.

A transitional approach uses LNG which reduces drastically emissions (85–95% for NO_x, 99% for SO_x, 91% for particles and 25% for CO_2). This would allow all ships to be retrofitted to use hydrogen. In 2015, the Carnival Corporation operator developed the first LNG-powered cruise ship. The world LNG fleet is rapidly increasing: in 2021, 250 units were operating and more than 400 ordered. Many programmes launched [8] have been successful (e4ships or HyFerry in Germany, SF BREEZE in the USA etc.).

For **submarines**, the German shipyards ThyssenKrupp Marine Systems units (Type 212A and Type 214) equipped with fuel cells using on-board stored hydrogen and oxygen are built in the 2000s. These submarines are operated by different countries like Greece, Turkey and South Korea. Norway and Germany have started developing an improved version in 2021, the Type 212CD for a first delivery in 2029. South Korea is developing a methanol reformer to produce hydrogen for fuel cell submarine as replacement of the hydride storage.

Having smaller sizes, many **vessel prototypes** using hydrogen have been developed or are in the planning phase. Fuel cells are used either for propulsion or auxiliary power supply. Applications cover cargo transport, tug boat, power supply in ports, offshore parks support etc. Between 2008 and 2013, a tourist boat was running in Hamburg under the Zemship programme. A dedicated hydrogen station allowed it to refill and store 50 kg for two PEMFCs of 48 kW each (100 kW propulsion engine). Some other projects or prototypes have been developed to show the possibility of usage of hydrogen/fuel cell for propulsion. Some research vessels use on-site hydrogen production to power the electrical propulsion motors. The "Race for Water" (e.g. "PlanetSolar") catamaran was equipped with 500 m^2 of solar modules, whose electricity is also used to produce hydrogen from seawater (5 kW electrolyser). It is stored in 25 cylinders at 350 bar (200 kg of hydrogen), which will be converted into electricity through two fuel cells of 30 kW each. The "Energy Observer", also a catamaran launched in 2017, uses electricity mainly from photovoltaic panels (130 m^2) to produce hydrogen (62 kg stored under 350 bar) and power the electric motors through a fuel cell. The French company EODev, a subsidiary of Energy Observer, has developed around Toyota's fuel cell hydrogen generators that could be used for boat propulsion or on-board generators.

River or sea ferries are also a target for fuel cell use. All American Marine, Inc. has launched the *Sea Change*, a 75-passenger ferry. The Cummins 360 kW fuel cell is fed by 242 kg of hydrogen providing a range of up to 300 nautical miles. The Norwegian operator Norled started operating the 82 m long, 300 passengers and 80 cars ferry *Hydra* (Figure 5.38). It uses liquid hydrogen to feed the Ballard fuel cell. Construction of other fuel cell ferries is planned in Norway, Spain.

Battery-powered vessels (especially ferries) have also been developed to reach zero emissions. Unlike hydrogen, no special "refuelling" infrastructure is required. The Norwegian Bastø Soren took into service in 2021 the 144 m long ferry *Bastø Electric* (Figure 5.39), at that point the largest electric ferry. It can carry 600 passengers, up to 200 cars and 24 trucks. Battery has a capacity of 4,000 kWh.

Figure 5.38: Hydrogen ferry *Hydra* (NORLED-Capman).

Figure 5.39: Electric/battery ferry *Bastø Electric* (Uavpic.com).

5.5.3.4 Aviation

Aviation represents about 2.4% of world CO_2 emissions. The use of alternative fuels (e-fuels) to reduce them is under consideration and a few tests were conducted. Hydrogen can be used with fuel cells with electric motors or as a direct fuel for turbo jets. However liquid hydrogen use for planes (gaseous is not an option even under 700 bar) is facing a critical issue: it has very low volumetric energy density (2.4 kWh/L) compared to kerosene (10.4 kWh/L). Liquid hydrogen cannot be stored in the wings as it requires super insulated cryogenic tanks. It would increase the operating cost compared to kerosene and, more critical, occupy a fairly large volume of the aircraft. First trials on test rig were conducted in 1937 in the US and in 1955, a Tupolev TU 155 flew with a turbojet powered by liquid hydrogen. A study has been carried out by Boeing in the mid-1970s on the conversion of a 747 for use of hydrogen instead of kerosene for reactors, while keeping the same characteristics (autonomy and number of passengers). If the quantity of fuel is reduced to 41 tonnes of hydrogen instead of 121 tonnes of kerosene, the volume required (578 m^3) exceeds any internal storage capacity. The project was then oriented towards two external reservoirs 4 m in diameter and 46 m long in the form of a nacelle under the wings. No concrete application followed. The European programme *Cryoplane* started in April 2000 was a feasibility study: the example of a small regional aircraft with 44 passengers and a range of 1,500 nautical miles required 30% of the cabin for the liquid hydrogen storage!

In 2021, Airbus presented three different *ZEROe* concepts (Figure 5.40) using turbofan or turbo prop as engine fed by liquid hydrogen (100–200 passengers and a range of 1,000–2,000 nautical miles). An A380 equipped with external prototype engine will be used as demonstrator in 2022. The European program Clean Aviation Joint Undertaking launched in 2022 a Call for Proposals. A study published in 2022 ("Hydrogen Powered Aviation") estimated the feasibility to have short range aircrafts using hydrogen by 2035. The 2021 UK FlyZero concept concerns a long-range (about 5,000 km) mid-size plane (279 passengers) using liquid hydrogen for the reactors. All those concepts avoid to emphasise the issue of the volume of the liquid hydrogen tank and also the psychological perception of passengers "sitting" on tonnes of liquid hydrogen.

Figure 5.40: Hydrogen *ZEROe* concepts and demonstrator *flightlab* (Airbus).

Another option is to use the hydrogen with a fuel cell powering electric motor/s ("all-electric plane"). Smaller planes with this approach are in a more advanced phase. In 2008, Boeing tested a two-seater demonstrator but did not develop it further. The

German research centre Deutsches Zentrum für Luft-und Raumfahrt (DLR) developed a first plane, the *Antares* (20 kW fuel cell and 42 kW electric motor) in 2009. It was followed by the HY4, a four-seater using a 45 kW fuel cell and batteries for a range up to 1,500 km with the 80 kW electric motor (Figure 5.41). It had two electric motors of 30 kW each. The US–British ZeroAvia has developed a hydrogen/electric "engine" that could be adapted to different planes, the Swedish company PowerCell supplying the fuel cell. After a prototype in 2020, the larger ZA600 unit will be tested on a 19-seater Dornier 228 plane (Figure 5.42) with 600 kW powertrain and range that should reach 500 miles. Certification is targeted for 2024. The US company Universal Hydrogen has developed a concept around modular capsules for transportation and storage of hydrogen (Figure 5.42). The first targeted certifications for 2025 are the 41-passenger aircraft De Haviland Dash-8 and the ATR 72. Fuel cells are supplied by Plug Power.

Electric planes powered by batteries or hybrid (ICE engine and electric motors) have been tested or are under development. The battery-powered demonstrator E-Fan from the Airbus Group made a first flight in 2014. The US company VoltAero is planning a five-seater hybrid aircraft (Cassio 330). The Eviation's Alice nine-seater prototype electric aircraft took its first flight in September 2022. It was powered by two 640 kW electric motors (820 kWh batteries). The series models will have a range of 440 nm (815 km). Certification is expected in 2023.

Figure 5.41: Hydrogen plane DLR Hy4 on top and first flight of Eviation battery plane (DLR, Jean Marie Urlacher and Eviation).

Figure 5.42: Zeroavia hydrogen plane (Zeroavia) and Universal Hydrogen regional aircraft with hydrogen capsule (Universal Hydrogen).

The use of hydrogen for commercial aviation as fuel for the turbojet or in a fuel cell is still in its infancy, even if the most optimistic developers are targeting a 2035 for a first flight. However, an hybrid approach (turboprop/electric motor) using either a fuel cell or batteries could be a transitional approach, allowing the reduction of emissions.

5.5.3.5 Other applications

One of the few sectors where the use of fuel cells/hydrogen is largely used is in the logistics sector. Since the first experiments in 1960, several thousand hydrogen **forklifts** are in operation. The economic sectors using them are large retailers (in 2022, Amazon, with 15,000 units, Walmart, USA, with 9,500 units; Carrefour, France; Lidl, Germany etc.) or factories (BMW, USA, with more than 400 units) or deliveries (FedEx). The players in this field are hydrogen suppliers like Linde Gas or fuel cells like Plug Power, HyGear or Nuvera in association with forklift manufacturers (Yale, Still, Hyundai, Toyota, Doosan Bobcat in South Korea or Heli in China). In 2022, Plug Power's fuel cells were used in 165 sites (over 50,000 units). The use of hydrogen, sometimes produced on-site, avoids the long charging times of the batteries and makes the forklifts able to operate continuously (Figure 5.43).

DAS BRENNSTOFFZELEN POWERPACK VON STILL:
Ein Komplettes Kraftwerk auf engstem Raum ohne CO₂-Emission

Figure 5.43: Hydrogen/fuel cell forklift (Still GmbH).

5.5.3.6 Hydrogen for internal combustion engine?

Hydrogen has already been considered to be used for ICE: the first ICE engine de Rivaz in 1806 was running using a mixture hydrogen/oxygen; BMW launched in 2005 its Hydrogen 7 car with a 6.0 L engine and 200 km range on hydrogen only and the Japanese company Mazda tested hydrogen on the RX-8 car with a Wankel rotary engine. Those two models were bi-fuel (gasoline could also be used). However, these vehicles are not zero emissions as nitrogen oxides NO_x are produced. Hydrogen ICE also need hardened (valves, valve seats etc.) or modified (air intake, fuel injection etc.) components. The maximum output is higher than gasoline or diesel engines by about 15%. Another advantage is the long experience and durability of ICE (no degradation of performance like fuel cell or batteries). This option could however be not commercially considered for cars, busses or trucks as many countries, especially in Europe, are planning to ban ICE car sales. However high power stationary engines require less hydrogen purity than fuel cells. Several institutes (FEV in Germany, Figure 5.44) or companies (MAN Energy Solutions or Cummins) are working on hydrogen-fuelled ICE. For the maritime sector or power generation, such an approach could be a way to reduce emissions. The Austria high power ICE supplier Innio Jenbacher (Figure 5.45) tested in 2021 different 12- and 16-cylinder engines (up to 1 MW_e) using up to 100% hydrogen or a mixture natural gas/hydrogen. Other series should be able to use up to 25% hydrogen starting 2022.

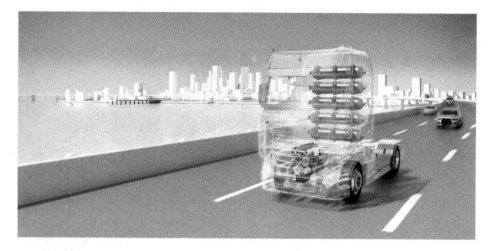

Figure 5.44: Hydrogen truck concept (FEV).

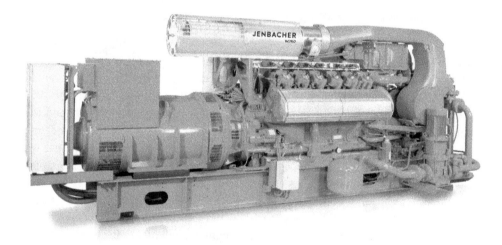

Figure 5.45: Stationary Hydrogen ICE engine Type 4 for stationary applications (Innio Jenbacher).

5.5.3.7 Hydrogen role in transportation

The different energetical issues associated to hydrogen like fairly low efficiency during production, compression or liquefaction, transportation and final use still leave some open questions for its use in road transportation (infrastructure or origin and cost of hydrogen, for example). In maritime application, the switch to hydrogen could be easier as more and more vessels will be LNG-powered. The space to store liquid hydrogen is also less critical as well as the refuelling infrastructure.

5.5.4 Domestic or commercial use

The domestic use of pure hydrogen is nowadays limited to a few experiments in areas such as micro-CHP. In Denmark, the first experiment carried out in 2008 on the island of Lolland validated the direct use of hydrogen to supply low-power domestic fuel cells by a distribution network of hydrogen produced by electrolysis (electricity from wind turbines).

In Germany, experiments at the level of a quarter "H_2-Communities" are conducted: they consist at least of an electrolyser and hydrogen storage. The city of Esslingen is testing this approach. The 1.7 MW_{peak} PV plant supply electricity to 1 MW electrolyser. Hydrogen is stored (30 kg/1,000 kWh) and used for mobility, industry and by a CHP unit supplying electricity and heat to 167 flats. Similar experiments are ongoing in the other German cities of Gütersloh, Aschaffenburg and Hamburg-Bergedorf.

In 2022, some manufacturers are proposing "H2Ready" prototype equipment like boilers or cookers that should be able to run with 100% hydrogen. The BDR Therma Group has started, after certification, evaluation of a domestic boiler in 2019 in Rozenburg, Netherlands. The UK programme Hy4Heat (2017–2022) explored the possibility to use hydrogen at a domestic level, including certifications and safety issues. Another approach is to run a low power fuel cell with hydrogen: it can then supply electricity and heat. Bosch and Ceres Power/UK are testing SOFC units with commercialisation targeted for 2024. However, in those units using a fuel cell, if electricity is the main production, heat is also generated (50–40% depending on fuel cell technology). There is no way to have a de-coupling, for example, when only electricity or only heat is needed or when too much heat is produced like in summer, for example.

An integrated approach for domestic use of hydrogen with solar PV, electrolyser, storage, fuel cell and batteries is proposed commercially by two companies. The German HPS Home Power Solutions AG with its PICEA can provide electricity and heat to a home. Measuring 1 × 2 × 1.75 m with 16 hydrogen tanks, it can supply 1.5 kW continuously (or 8 kW for 3 h) of electricity. Heat from the 1.5 kW fuel cell (hydrogen supplied by an electrolyser with a capacity of 500 L/h) can be stored. PV requirement is about 17 m². The Australian Company LAVO system uses hydrides to store (up to 40 kWh) the hydrogen generated by an electrolyser. The unit (Figure 5.46) is compact: 1.68 × 1.24 × 0.4 m and weighs 324 kg. Those two equipments are still expensive (€60,000–90,000 for the PICEA in 2022) to be used on a large scale.

5.6 Hydrogen as a commodity: import/export

Hydrogen is already imported by some countries but in relatively small quantities. China is the largest importer with 25 million tonnes in 2021, the USA being the second with 320 million m³, followed by Netherlands and Germany. But the availability of large quantities of renewable energies (wind or sun) in some countries (Saudi Arabia, Oman,

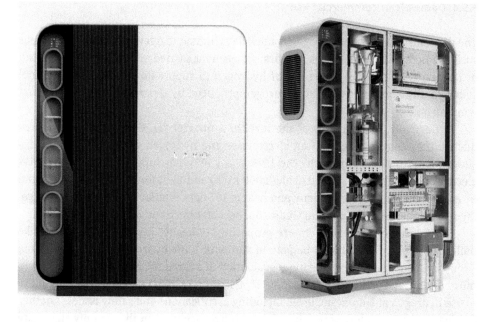

Figure 5.46: Integrated hydrogen domestic "power plant" (LAVO System).

North Africa, Chile, Norway, even Ukraine etc.) have led to another approach to hydrogen strategy: import from countries where electricity price is low and to minimise investments costs for the importers. Numerous partnerships for green or low CO_2 hydrogen have been either initiated or signed. Main importers are Germany, Japan and South Korea. Contracts have been signed with Australia (for Germany and Japan) and Portugal (for Netherlands). Netherlands is building an infrastructure in Rotterdam not only to produce hydrogen on-site but also to receive hydrogen carriers at an import terminal (operational in 2026) and deliver the hydrogen to Germany through pipelines (up to 4.6 million tonnes by 2030). The German energy supplier E.on and the Australian Fortescue Future Industries have signed an agreement for hydrogen supply to Germany in 2022. From 2024 to 2030, 5 million tonnes of green hydrogen will be delivered.

5.6.1 Low CO_2 hydrogen: Australia

Australia, in a project with Japanese companies led by Kawasaki, has developed hydrogen production from brown coal. Up to 225,000 tonnes of hydrogen are expected to be produced by 2030. The classification as low CO_2 hydrogen is due to the CCS technology. The low coal prices make the hydrogen production economically viable. However, the liquefaction and transport through long distances (to Japan or the Netherlands) are not in line with energy efficiency.

5.6.2 Limitations

The hydrogen "thirst" of some European countries leads to some contradictory strategies: on one hand, decarbonise society (industry, energy sector, transports etc.) with strategies oriented towards efficiency and on the other hand accept worldwide hydrogen decentralised production and long distance transportation. Hydrogen carriers are just at the starting phase. The first vessel (Suiso Frontier) was launched in 2019. The first delivery of 88.56 tonnes of Australian LH2 to Japan was achieved in 2022 after a 23-day trip. Considering the expected exported volumes, a huge fleet of carriers will be needed.

The COVID pandemic and the war in Ukraine have shown the fragility of such structures and the danger of depending too much on just a few countries for energy or materials. Some targeted countries (North Africa, for example) are also faced to low water availability. Thus, water desalination would have to be build, adding cost to the hydrogen.

The environmental advantage of green hydrogen is the local (in a country or a region) production and use.

5.7 Hydrogen, ammonia or methanol?

If hydrogen is the first step of the power-to-gas, its use is still limited either due to the specific properties (low gaseous density, low temperature of LH2, energy needed to compress or liquefy it and so on).

Why not converting it into another form that could be easier to handle?

5.7.1 Ammonia

Ammonia is the first answer. It is one of the most produced industrially worldwide (150 million tonnes in 2021). It is not only used primarily for fertilisers but also as refrigerant gas for plastic or explosive manufacturing and so on

The basic reaction is

$$N_2 + 3H_2 \Leftrightarrow 2NH_3.$$

Process efficiency is about 65% starting form hydrogen produced through steam methane reforming or 55% if electrolysis is used to produce hydrogen.

At room temperature, ammonia is a highly irritating and corrosive colourless gas. Its density under STP is 0.769 kg/m^3. Liquefaction temperature is –33 °C at normal temperature or 20 °C under 7.5 bar (17.8% weight hydrogen content) thus making transportation easier than hydrogen. Its energy density is 21.8 MJ/kg or 12.7 MJ/L (4.32 kWh/L). Handling and safety in industrial environments are well documented and applied by skilled workforce. Main issue are leaks in pipes or tanks. Recent recorded accidents were in 2009 in a fertiliser factory in the USA and in 2016 in Bangladesh also in a fertiliser company.

Transportation of ammonia is a daily operation in the more than 120 ports worldwide having the appropriate infrastructure. Large vessels ammonia carrier have a capacity reaching 90,000 m^3. The US company Nustar delivers ammonia since 1971 through a 3,000 km pipeline network carrying 1.5 million tonne/year.

Ammonia as hydrogen carrier can provide a more convenient solution to hydrogen transport and delivery. The first approach involves cracking ammonia in a dedicated site and then delivering the hydrogen (trucks, pipelines) to users (industry, refuelling stations etc.). The second one is using an on-site cracker followed either by direct use or by compression of hydrogen. For long distance hydrogen delivery (Australia–Japan, for example), ammonia is a convenient way as it does not require the very low temperature needed for LH2. After cracking, the only by-product is nitrogen. Another option to use indirectly ammonia is to crack it and produce hydrogen that can be directly used in a fuel cell or ICE engine. In 2022, a John Deere 100 kW mid-size tractor was converted by the Amogy Inc. company using ammonia cracking modules with a fuel cell using the hydrogen produced.

Ammonia as potential fuel for ammonia vessel carriers is an option investigated, especially considering the challenges associated: safety issues, on-board storage, regulation and validation. The Japanese NYK Line, Nihon Shipyard, ClassNK and the Norwegian chemical company Yara International have started a study involving an ammonia-powered engine in 2021. The German company Man Energy Solutions is targeting a two-stroke ammonia engine for 2024/2025. The Finnish Wärtsilä has already tested a four-stroke demonstrator running on ammonia in 2022.

Besides ICE engines, SOFC working at high temperature (>550 °C) can use ammonia directly to produce electricity and heat and has the same power/current profile as using hydrogen or natural gas and showing even a slightly higher efficiency.

5.7.2 Methanol

Methanol is another option. At room temperature, methanol is liquid making its use easier than ammonia. It is also less toxic. In 2021, world production was about 110 million tonnes. Methanol is used as chemicals feedstock (ink, resins, adhesives etc.) and for transportation (gasoline additive), with a density of 0.796 kg/L and an energy density of 15.8 MJ/L (20 MJ/kg; 5.5 kWh/kg – LHV).

It is produced by direct synthesis using syngas ($H_2/CO/CO_2$), the global reaction being

$$CO_2 + 3\,H_2 \Leftrightarrow CH_3OH + H_2O.$$

Methanol as hydrogen carrier with its hydrogen content of 12.6% (17.8% for ammonia) can be an alternative to the direct transport of hydrogen: a 40 tonnes truck carrying hydrogen under 200 bar can supply 500 kg of hydrogen; a 40-tonne methanol truck can supply the equivalent 3,600 kg of hydrogen under normal conditions (pressure and temperature). Long distance methanol transport cost could be equivalent to LH2 delivery [9].

The interest for methanol as fuel for vessels is increasing: in 2021, Maersk announced that by 2024, eight vessels (16,000 containers 20 feet equivalent) will be fuelled by methanol. Maersk has also secured the supply of green methanol. Since 2015, the StenaLine's ferry "Germanica", built in 2001, uses a methanol-powered Wärtsilä engine of 32,000 HP. Shipping companies are moving to methanol or dual fuel vessels: Maersk/Denmark, AIDAnova/Germany, Sumitomo Heavy/Japan, Port of Antwerp (tugboat), etc. Numerous companies (ship design, engine manufacturers, shipyards etc.) are developing the use of methanol. Methanol engines are also developed as dual fuel (MAN EnergySolutions since 2012). Retrofits are also available for existing engines. *Scandi*NAOS AB/Sweden, for example, is proposing methanol engines up to a power of 450 kW.

Compared to ammonia, there is no need of a heat source to decompose ammonia into hydrogen. Compared to LNG, methanol does not require the low temperature (−162 °C) and has roughly the same energy per volume.

Methanol can be used directly by low temperature fuel cells (DMFC). In the fuel cell, at the anode, the mixture methanol/water is decomposed into CO_2 and H^+ ions thus liberating electrons. The suppliers are SFC Energy AG/Germany with low power units, Blue World Technologies/Denmark planning up 25 kW power DMFC etc.

5.7.3 Fuel comparison

The main energetic and volumetric characteristics for the hydrogen, ammonia, methanol and LOHC are summarised in Table 5.7.

LOHC (see Section 5.3.6.1) can also be another solution for hydrogen transportation although it needs a specific infrastructure.

The other issue in this comparison is the cost. The war in Ukraine beginning 2022 has completely shaken the different markets (raw material, electricity, transportation etc.) thus making a mid or long-term comparisons unreliable.

If the electrolysis makes possible to recover the surplus of electricity of renewable origin, the relatively low overall energy efficiency and the infrastructure needed must lead first to the direct use of this electricity. The different fields of application

(industry, energy and transport) must be further optimised in terms of energy efficiency and equipment.

Table 5.7: Comparison of different hydrogen alternatives.

	LH2	LNG	Ammonia	Methanol	LOHC
Boiling temperature	−253 °C	−162 °C	−33 °C	65 °C	>300 °C
Density	70.8 kg/m^3	460 kg/m^3	682 kg/m^3	795 kg/m^3	1050 kg/m^3
Gravimetric energy density LHV (MJ/kg)	120	49	18.6	20.3	
Volumetric energy density (MJ/L)	8.52	22.2	12.8	15.7	
Hydrogen content	100%	≤25 wt%	17.8 wt%	12.5 wt%	57 kg/m^3
CO$_2$ emissions	0	22.4 L/mol CH$_4$ (16 g)	0 (N$_2$, NO$_x$ emissions)	22.4 L/mol (32 g)	

However, power-to-gas technology is the only way to store the large expected surpluses. The hydrogen produced and methane from methanation contributes to overall
energy efficiency, the first step in an energy transition.

There are still many technological barriers, but the projects and achievements
make it possible to bring this technology to large-scale commercialisation.

References

[1]   Robinius, M. Strom– und Gasmarktdesign zur Versorgung des deutschen Straßenverkehrs mit
      Wasserstoff. Fakultät für Maschinenwesen der Rheinisch–Westfälische Technischen Hochschule
      Aachen. Thesis, 2015.
[2]   European Hydrogen Backbone, A European Hydrogen Infrastructure Vision Covering 28 Countries,
      April 2022.
[3]   Liemberger, W. Extraction of Green Hydrogen at Fuel Cell Quality from Mixtures with Natural Gas.
      Chemical Engineering Transactions, p. 427–431, 52 (2016).
[4]   Tegel, M, et al. An Efficient Hydrolysis of MgH$_2$-Based Materials. International Journal of Hydrogen
      Energy, 42, 4 (2017), 2167–2176.
[5]   Bossel, U. The Birth of the Fuel Cell 1835–1845. 2000. European Fuel Cell Forum.
[6]   ADEME/IFPEN, "TRANPLHYN" Transports lourds fonctionnant à l'hydrogène, 2022.
[7]   VDI/VDE, Klimafreundliche Nutzfahrzeuge, 2022.
[8]   EMSA European Maritime Safety Agency. Study on the Use of Fuel Cell in Shipping, 2017. DNV GL
      Maritime.
[9]   Schorn, F, Breuer, JL, Samsun, RC, Schnorbus, T, Heuser, B, Peters, R and Stolten, D. Methanol as a
      Renewable Energy Carrier. Advances in Applied Energy, p. 1–14, 3 (2021).

6 Beyond power-to-gas

6.1 Power-to-liquid

6.1.1 Synthetic chemicals and fuels

In the 1920s, German chemists Franz Fischer and Hans Tropsch were able to synthesise the first hydrocarbons according to the exothermic reaction called Fischer–Tropsch (FT), e.g. for alkanes:

$$(2n+1) \ H_2 + nCO \Rightarrow C_nH_{(2n+2)} + nH_2O \tag{6.1}$$

Starting materials were hydrogen and carbon monoxide (syngas) from coke gas. A catalyst (iron, cobalt or ruthenium) was required, and the reaction was carried out under high pressure and temperature. The choice of conditions (composition of the mixture, catalyst, temperature, pressure etc.) makes it possible to obtain different hydrocarbons:

$$2CO + 4H_2 \rightarrow C_2H_5OH + H_2O \ (\text{methanol}) \tag{6.2}$$

$$8CO + 17H_2 \rightarrow C_8H_{18} + 8H_2O \ (\text{octane}) \tag{6.3}$$

As a result of the ever-increasing discoveries of oil and natural gas deposits and the corresponding decline in prices, this approach has not been as successful as expected. It was, however, used by countries without oil or gas resources (Germany between the mid-1930s and the Second World War to produce synthetic fuels) or without access to the fuel market (South Africa during the boycott following apartheid).

However, surplus electricity from renewable sources in the coming decades will increase the availability of hydrogen in significant quantities and may allow a renewal of this technology to produce synthetic fuels or chemicals.

The term "synthetic" can be misleading: it does not mean that the fuels are produced from basic atoms (H, C and O) but from chemical conversion.

6.1.2 Power-to-liquid

To produce the initial mixture (syngas) used by the FT reaction, the power-to-gas supplies hydrogen, whereas carbon dioxide (CO_2) can be derived from a biogas unit or captured from the air. CO_2 is reduced to carbon monoxide (CO) using oxygen from the air. High-temperature electrolysis provides the optimum conditions for the use of excess electricity to the production of hydrocarbons (Figure 6.1): steam electrolysis makes it possible to increase the efficiency of electrolysis and of the conversion to hydrocarbon.

https://doi.org/10.1515/9783110781892-007

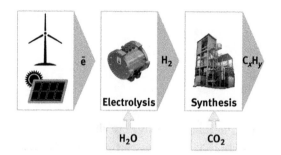

Figure 6.1: Power-to-liquid principle.

6.1.3 Gas-to-liquid

In this approach, **natural gas** is used as the starting point to produce synthetic compounds. The first step is the production of syngas according to the reactions

$$CH_4 + H_2O \Rightarrow CO + 3H_2 \text{ with } \Delta H^0 = 206 \text{ kJ/mol} \tag{6.4}$$

$$CH_4 + CO_2 \Rightarrow 2CO + 2H_2 \text{ with } \Delta H^0 = 247 \text{ kJ/mol} \tag{6.5}$$

Two important reactions are the exothermic water gas shift reaction, which converts CO into hydrogen and reverse water gas shift. The two coexist in an equilibrium varying with the temperature:

$$CO + H_2O \Leftrightarrow CO_2 + H_2 \text{ with } \Delta H^0 = -41 \text{ kJ/mol} \tag{6.6}$$

The synthesis of methanol, for example, occurs according to the reaction:

$$CO_2 + 3H_2 \Rightarrow CH_3OH + H_2O \text{ with } \Delta H^0 = -62 \text{ kJ/mol} \tag{6.7}$$

Experiments are mainly directed towards the production of synthetic fuels.

6.1.4 PtL experimentations

6.1.4.1 Production of synthetic diesel

The pilot plant initiated by Audi and carried out by Sunfire (Figure 6.2) was built in Werlte and has been operational since the end of 2014. The unit produces synthetic diesel (Figure 6.3).

An SOEC high-temperature electrolyser (10 bar and 800 °C) supplied by Sunfire dissociates the water into hydrogen and oxygen with 90% yield.

Hydrogen reacts with CO_2 from a biogas unit in a reactor under pressure and temperature to produce a liquid hydrocarbon (*Blue Crude*). The efficiency of the operation is of the order of 70%. By distillation, this hydrocarbon can be converted into synthetic

POWER-TO-LIQUIDS

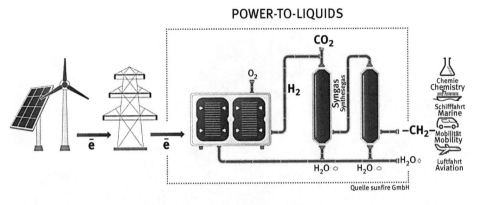

Figure 6.2: Power-to-liquid with high-temperature electrolysis (Sunfire).

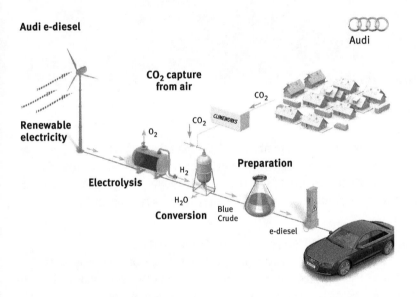

Figure 6.3: Diagram of the power-to-liquid unit in Werlte (Audi).

diesel (160 L/day with 80% yield) without sulphur or aromatics and with a high cetane number (Figure 6.4).

A similar plant is to be installed in Norway (Heroya Industrial Park) in 2022 with a production start-up in 2024. Nordic Blue Crude, Sunfire (high-temperature electrolysis), Climeworks (capture of air CO_2) and EDL Anlagenbau (infrastructure) companies are involved in this project with an initial capacity of 10 million L/year. A 2022 project in Portugal near Porto will use CO_2 (up to 100,000 tonnes/year) from waste gas and hydrogen to produce e-kerosene. In Austria, AVL List GmbH should have started in 2022 in Graz, a unit with a yearly capacity of 100,000 tonnes of e-fuel. The Norwegian Norsk e-Fuel plans a unit in northern Norway with a capacity of 25,000 m^3 to be

Figure 6.4: Power-to-liquid unit in Werlte (Audi).

operational in 2026 with an increase to about 100 million L in 2029. The French energy supplier Engie is planning an e-fuel unit in Dunkirk, north of France. A 400 MW of electrolyser capacity and 300,000 tonnes of CO_2 from ArcelorMittal steel plant would allow to produce up to 100,000 tonnes of e-fuels, which should be operational in 2026.

6.1.4.2 MefCO$_2$ project (2014–2019): production of methanol

An installation near Dortmund in Germany produces methanol from hydrogen obtained by electrolysis (1 MW Hydrogenics electrolyser producing 200 Nm^3/h of hydrogen). CO_2 comes from the Steag coal-fired power plant. The methanol production unit was developed by Carbon Recycling International. Production which began in 2017 with a capacity of 250,000 L/year.

6.1.4.3 Soletair project

This research project was carried out by the Finnish Research Centre (capture of CO_2 from air), the Finnish University of Lappeenranta (electrolyser) and the German Institute of Technology KIT – Karlsruhe Institute of Technology (compact synthesis reactor developed by the spin-of INERATEC – Figure 6.5).

The initial equipment allowed to validate the operation of the unit installed in Finland (Figure 6.5), which uses the electricity of a photovoltaic plant. Its capacity is 80 L of gasoline per day, and in July 2017, nearly 200 L was produced after the first start.

Figure 6.5: Synthetic reactor (INERATEC) and power-to-gas research unit in Finland (VTT Technical Research Centre of Finland Ltd).

The German Fraunhofer Institute has been carrying out its project "Strom als Rohstoff" (electricity as a raw material) since 2015. Various demonstrators aim at the production of hydrogen peroxide (H_2O_2), ethylene (C_2H_4) or alcohols (C_1–C_{20}). Objectives of the project are to demonstrate the feasibility of these PtL options.

In 2022, INERATEC GmbH supplied to atmosfair GmbH (an organisation that allows travellers to compensate CO_2 emissions) a plant (Figure 6.6) that could produce up to 350 tonnes of synthetic crude oil that can be refined as kerosene. CO_2 is captured from air (1,000 tonnes/year). Another project from INERATEC is for the chemical company Höchst with a capacity of 3,500 tonnes/year. In 2022, a partnership with the Japanese engineering company Chiyoda Corporation allowed the development of INERATEC's technology in Japan.

Figure 6.6: e-Fuel production plant in container (INERATEC).

6.1.4.4 Uses of e-fuel

If the chemical industry could use the power-to-liquid products, for transportation (cars, trucks, busses but especially aviation) the potential advantages are subject to questioning. The first issue is the productivity of the whole chain: electrolysis, transportation (compression) and conversion to liquid lead to well-to-wheel of less than 20%. The second issue is the price of the e-fuel: depending on the source, the projections are very variable. In 2022, e-kerosene is estimated to be 6 to 8 times more expensive than kerosene. The average estimations see the price of e-fuel to decrease to €2–2.5 per litre in 2030–2050. Will there be enough e-fuel considering the current consumptions as only small production units are running? According to the IEA, kerosene consumption in 2021 was about 57 billion gallons (216 million m^3). Replacing only 10% of kerosene by sustainable aviation fuel would require about 21 million m^3 of e-fuel. The Norwegian Norsk e-Fuel is planning a plant for production in 2024. It will have a capacity of 25,000 m^3 in 2026 and 100,000 m^3 in 2029. Car manufacturers are also pushing the use of e-fuel to keep internal combustion engine (ICE) running cleanly. A German-Chilean consortium with Porsche and Siemens has started in 2021, building an e-fuel plant in Chile. The production target was 130 m^3 in 2022, expanding to 55,000 m^3 in 2024 and 550,000 m^3 in 2026. As a comparison, the German gasoline consumption in 2021 was 21 million m^3 (42 million m^3 diesel).

The future for e-fuels [1] as a replacement of conventional fuels still has many challenges to overcome, the most important being cost and volumes that currently and, even according to the planned investments until 2030, will represent a minute fraction of the global needs. The other issue is the global yield: for transportation, the wheel-to-wheel for battery vehicles is about 80%; for hydrogen and fuel cell vehicles, it is only 25–30% but for ICE using e-fuels it is at best 20% or less when imported from non-European countries.

6.1.4.5 Conclusion

Power-to-gas technology opens up a new perspective to the production of synthetic fuels or chemical compounds of high purity and is an alternative to fossil fuels (petroleum, natural gas and coal). The reached yield makes it possible to envisage commercial production.

Another advantage is the low greenhouse gas emissions for PtL using renewable electricity, CO_2 and water compared with conventional oil or other chemicals obtained from natural gas, oil or coal.

CO_2 from industrial processes or power plants should reduce the necessity to move to carbon capture and storage.

6.2 Power-to-heat

6.2.1 Principle

Another option for storing excess electricity is the conversion into heat stored in tanks or other materials, at the domestic, urban or industrial level. This option could be an alternative to supplement battery storage where no power-to-gas unit is available to absorb the excess electricity when the battery is fully loaded (Figure 6.7).

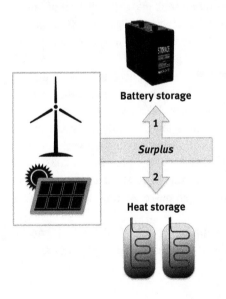

Figure 6.7: Principle of the power-to-heat concept.

However, the conversion of electricity to heat should only be activated when electricity generation is in excess and cannot be used otherwise.

6.2.2 Heat storage

Apart from water, other materials can be used to store heat from the conversion of excess electricity: phase change materials, stone, concrete etc.

The Dutch company Ecovat BV has built in Uden a prototype of underground storage of heat for an office building. The reservoir (20 m in diameter and 26 m in height) has a capacity of 1,500 m^3 of water (88,000 kWh). The insulation keeps hot water (>90 °C) for several months with less than 10% losses in 6 months.

"Thermal batteries" use a phase-change salt instead of water. The German company HM Heizkörper developed a hybrid water/sodium acetate trihydrate module (40 cm diameter for 1.8 m high) capable of storing a latent heat of 2 kWh over a long period (several weeks or months). However, as of 2022, the company had no more activity in this area. The Norwegian research organisation SINTEF Energy installed a bio-wax thermal battery in a zero-emission building (ZEB Lab) in Trondheim in 2021. The 3 tonnes (5 m^3) of bio-based wax can store up to 200 kWh of heat allowing up to 2–3 days of autonomy (Figure 6.8).

6.2.3 Urban experiments

In **Germany**, since 2015, a large-scale experimentation took place in a district of Berlin (Adlershof). A central combination of cogeneration using natural gas (96 MW$_{th.}$ and 13 MW$_{él.}$) and storage of heat makes it possible to regulate the supply of electricity (Figure 6.9).

In case of electricity deficit, the power of the CHP (combined heat and power) units is increased, and in case of excess electricity, it is converted into heat and stored in five tanks totalling 2,000 m^3 of water (6 MW) for distribution in the local district heating network.

Also in Berlin, the energy supplier Vattenfall is investing from 2017 in a unit capable of storing 120 MW$_{th.}$, which will also supply the local district heating network. In 2022, another storage tank with a capacity of 56,000 m^3 (2,600 MWh) is planned for 2023.

In **Switzerland**, since 2016, Alpiq installed two units of 11 MW each in Lausanne to produce steam from surplus electricity at the Gösgen power station. This steam was used by a paper mill.

The German company ELWA provides PtH units up to 1.67 MW, which can be used among domestic, commercial, and excess electricity from CHP applications.

6.2.4 Experiments in industry or agriculture

Manufacturers requiring hot water or steam production can also use power-to-heat (PtH) technology to reduce production costs. This approach can be done internally or through contracting.

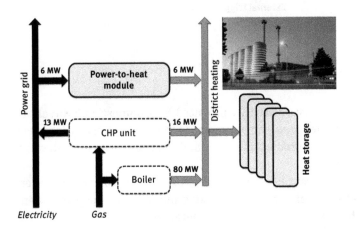

Figure 6.8: SINTEF ZEB Lab with bio-wax storage tank (SINTEF).

Figure 6.9: Principle of operation of the power-to-heat Adlershof concept (data: BTB GmbH, Berlin).

In this context, the German sugar producer Südzucker has installed a 10 MW PtH unit in 2016, both for internal needs and to supply a local steam network, thus increasing the flexibility of the installation. The Norwegian company PARAT Halvorsen AS installed in 2022 in Finland is a 30 MW unit of a greenhouse to produce cucumbers. The compact units are available up to 60 MW power and provide hot water and steam.

6.2.5 Domestic experiments

The direct use of excess electricity uses the resistances of existing **hot water tanks**, which can be controlled by the electricity supplier and activated only in the event of a surplus (Figure 6.10).

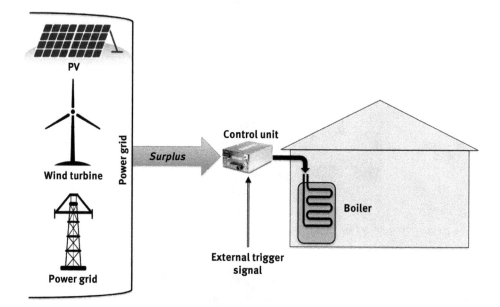

Figure 6.10: Domestic power-to-heat concept.

The other option (Figure 6.11) uses electricity produced by a **domestic photovoltaic system**. Many companies offer management modules that can optionally integrate battery storage: they detect overproduction phases and send electricity to the heating resistance of the hot water tank.

Heat pumps can also use surplus electricity from renewable sources. However, the real efficiency (coefficient of performance taking into account the primary energy) is still low as the heat pumps are not used as energy storage.

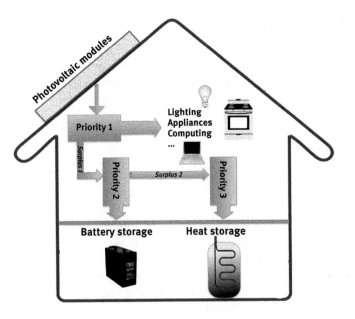

Figure 6.11: Power-to-heat from photovoltaic installation.

An energetically limited approach

The transformation of electricity into a lower form of energy (heat) leads to an energy "deadlock" because low-temperature heat can hardly be converted back into another form of energy or with a sufficient yield. However, the PtH approach can be a complement to maintain the stability of the electricity grid or use the excess electricity produced locally that can no longer be stored or converted into another form.

6.3 Combination of PtH and PtL

For the industry, PtH technology has a very short reaction time and can almost imme-diately use excess electricity. The PtL process requires a start-up time to have water or steam at a given temperature. The combination of the two approaches is based on the use of excess electricity to first power the PtH unit to immediately use this surplus and then to allow the PtL unit to take over with steam at the required temperature.

Reference

[1]   Power-to-Liquids, A Scalable and Sustainable Fuel Supply Perspective for Aviation, German Environmental Agency (Umwelt Bundesamt), January 2022.

7 Power-to-gas experiments

While one of the earliest experiments took place at the end of the nineteenth century by the Dane Poul la Cour (see Chapter 3), more recent research projects were launched in the mid-1980s, followed by projects with higher power pilot units.

7.1 Early developments

7.1.1 HySolar programme (1986–1995)

In the mid-1980s, a 10-year research contract was signed between Saudi Arabia and Germany for the HySolar (*Hydrogen from Solar Energy*) programme. This programme saw the design and installation, between 1985 and 1989, of a building in Stuttgart (Figure 7.1) housing a pilot unit of 10 kW. It used electricity supplied by photovoltaic panels (14.7 kW) to produce hydrogen (2 kW electrolyser with an efficiency of 80%) which was compressed at 200 bar and stored in tanks. The hydrogen produced has been used for research on catalytic combustion, for fuel cells (alkaline fuel cell (AFC) or phosphoric acid fuel cell (PAFC) type) or for modified internal combustion engines.

A 2 kW teaching and research unit (photovoltaic panels and electrolyser) was installed on the rooftops of the University of Ryad. A demonstrator using electricity from photovoltaic panels (3,800 m^2) was built north of Ryad and commissioned in 1993. It consisted of a 350 kW electrolyser supplying compressed hydrogen at 150 bar stored in tanks.

HySolar: a precursor programme

The approach of this programme has laid the foundation for a power-to-gas (P2G) installation. The numerous visitors (up to 3,000 per year) were able to judge the validity of this concept despite the relatively low yields linked to the technological development of that time. The objectives and concerns corresponded to those of the current facilities.

Other experiments were carried out in the 1990s. One of them, led by the Spanish institute Instituto Nacional de Técnica Aeroespacial, included the entire energy chain:

- Photovoltaic panels (144 modules) producing 7.5 kW
- An alkaline electrolyser of 5.2 kW (1 Nm^3/h of hydrogen)
- A purification system followed by storage of hydrogen combining metal hydrides and reservoir under a pressure of 6 or 200 bar
- A 10 kW PAFC dual fuel cell capable of using either hydrogen or methanol with a reformer

https://doi.org/10.1515/9783110781892-008

Figure 7.1: HySolar building in Stuttgart (DLR – Deutsches Zentrum für Luft- und Raumfahrt).

7.1.2 Projects at the level of a building

In the programme run between 1992 and 1995 by the German institute Fraunhofer in Freiburg, a low-energy building was specially designed for this concept. Electricity consumption has been minimised by the use of energy-saving equipment, which included:
- photovoltaic panels (34 m² producing 4500 kWh/year),
- lead-acid batteries for electricity storage with a capacity of 19.2 kWh,
- a proton exchange membrane (PEM)-type electrolyser of 2 kW producing hydrogen under a pressure of 30 bar,
- external storage of hydrogen (15 m³) and oxygen (7.5 m³) and
- a fuel cell of the PEM type.

Hydrogen was used for heating by catalytic combustion without flame. It was also used for cooking. When the production of photovoltaic electricity was insufficient or when the batteries were discharged, a fuel cell took over for the production of electricity. It operated at a temperature of 70 °C with an electrical efficiency of 60%. The heat produced was recovered to heat water.

Solar energy was also used for the thermal part with 12 m² of collectors. After experimentation and dismantling of the equipment, the building still houses offices.

On an even smaller scale, the Swiss Markus Friedli equipped his home in the 1990s with a system combining solar and hydrogen. Photovoltaic panels with a peak power of 7.4 kW powered an alkaline electrolyser of 10 kW. The hydrogen produced (0.3–0.95 Nm³/

h) is purified (removal of the carried potash, oxygen and water), compressed and stored in a buffer tank and then in hydrides (91 L tank weighing 235 kg and containing up to 19 Nm^3 of hydrogen). Hydrogen is used for a stove and a modified minibus which also stores hydrogen in hydrides.

Other experiments to use hydrogen produced from wind or photovoltaic energy at a housing level have been conducted since the 1980s.

7.1.3 Exploratory projects

Exploratory projects were carried out in many countries in the early 2000s. These include:
- in Great Britain between 2000 and 2004 (PURE project on the island of Unst),
- in the USA in 2001 in Reno and in 2004 in Chicago,
- in Canada in 2003,
- in Italy in 1997 and 2000 and
- in Germany in 2003 (project PHOEBUS).

Most of these projects used a PEM fuel cell (PEMFC) with a few kilowatts of power (typically <10 kW) to generate electricity from stored hydrogen.

7.1.4 First field experiments

7.1.4.1 Solar Park/hydrogen from Neunburg vorm Wald (Germany)
This project, launched in 1987, was based on a photovoltaic plant of 350 kW associated with three electrolysers of about 100 kW each. The hydrogen produced was stored in gaseous (tanks of 6,000 m^3 under 30 bar) or liquid form (tanks of 3,000 L).

Hydrogen gas, mixed with natural gas, was supplied to several boilers or was converted into electricity by three fuel cells (one mounted on a forklift and the other two stationaries, respectively, of the AFC type of 7 and 79 kW). A liquid hydrogen service station, produced on another site, was used to test the filling of tanks of modified ICEs (internal combustion engines) supplied by BMW and running on hydrogen.

7.1.4.2 Utsira (Norway)
One of the first complete installations was that of the Norwegian island of Utsira and experimented between 2004 and 2008. On this island of 220 inhabitants, 2 wind turbines of 600 kW each were put into service in 2003. In order to be able to store electricity in excess, an installation comprising an electrolyser, a hydrogen compressor and a reservoir for its storage has been used. This hydrogen supplied a fuel cell and an ICE generator that provided electricity uninterruptedly (up to 48 h without wind)

Figure 7.2: Power-to-gas installation on the island of Utsira (Norsk Hydro ASA).

to 10 houses on the island. A flywheel regulated electricity supplied directly to the grid by wind turbines (Figure 7.2).

The technical characteristics of the installation are summarised in Table 7.1.

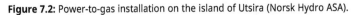

Table 7.1: Utsira power-to-gas components.

Equipment			
Wind turbines	2 × 600 kW	Blade diameter 40 m	Enercon
Electrolyser	10 Nm³		Norsk Hydro
H₂ storage	2400 L to 200 bar		Norsk Hydro
Fuel cell	10 kW	PEMFC	IRD
Generator	55 kW		Continental
Flywheel	5 kW		Enercon

The final electrical yield was of the order of 25% but allowed to continually feed the houses, which is an important point on an island. At the end of the experimentation, only the wind turbines were kept.

Early promising approaches

All these initial experiments validated the P2G technology despite the technological limitations mainly related to equipment that was not always optimised or that had low yield. Projects over a long period have also shown economic and social viability (e.g., experimentation on islands), allowing an uninterrupted supply of electricity.

7.2 Research projects

Following the exploratory programmes of this technology from the 1980s to the 2000s, other institutes, companies and organisations have launched higher power units, bringing together the technological advances of the last few years (e.g. methanation).

7.2.1 The ZSW Institute in Stuttgart

In collaboration with the Fraunhofer Institute IWES, the German research centre for solar and hydrogen ZSW (Zentrum für Sonnenenergie– und Wasserstoff-Forschung) near Stuttgart and the ETOGAS company (formerly Solarfuel GmbH, bought in 2017 by the Swiss company Hitachi Zosen Inova) evaluated a high-power demonstrator of 250 kW$_e$ (Figure 7.3).

Figure 7.3: ZSW driver installation (ZSW-BW/Solar Consulting GmbH).

The installation, which follows a low-power unit launched 3 years before, consisted of:
–   an alkaline electrolyser (Figure 7.4) with a power of 295 kW providing up to 65 Nm^3 of hydrogen per hour (or 300 m^3/day) under a pressure of 6–11 bar and
–   a methanation plant with a methane capacity of up to 15 Nm^3/h

Figure 7.4: Alkaline electrolyser (ZSW-BW/Solar Consulting GmbH).

The results and experiences of this project have been used for other pilot installations (e.g. Audi "e-gas").

7.2.2 Project WIND2H2

The US WIND2H2 project led by the National Renewable Energy Laboratory (NREL) and the XcelEnergy energy provider between 2007 and 2010 explored various options (Figure 7.5).

Electricity was supplied by a 10 kW photovoltaic plant and two wind turbines (10 and 100 kW). The hydrogen produced either by two PEM electrolysers of 6 kW (2.25 kg H_2/day) or by an alkaline electrolyser of 33 kW (12 kg H_2/day) was compressed under 240 bar and stored (115 kg capacity). This hydrogen made it possible to feed:

– a 5 kW fuel cell,
– a modified 60 kW ICE engine and a generator (P2G2P) and
– a service station, after compression to 400 bar and storage (capacity of 18 kg).

This programme allowed to confirm the better efficiency of the PEM electrolyser (57% versus 41% for alkaline) and to extrapolate the 2016 cost of hydrogen to US $5.83 for a larger 2.33 MW electrolyser.

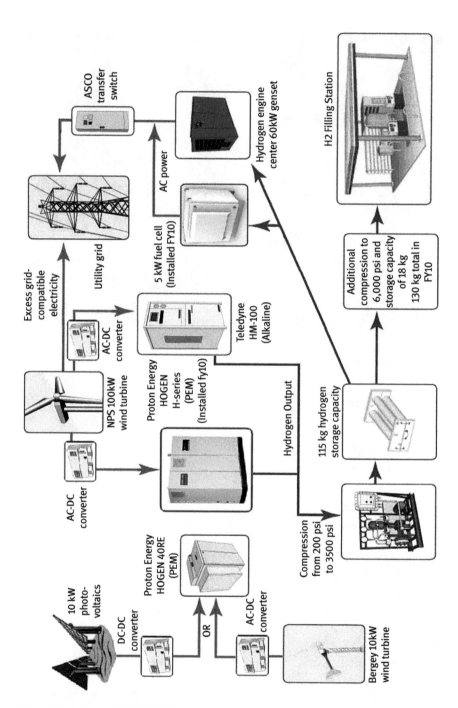

Figure 7.5: Project WIND2H2 (NREL).

7.2.3 National programmes or international cooperation

Many national or European programmes have been or are being carried out in various fields related to the P2G concept. They mainly concern hydrogen.

7.2.3.1 Examples of projects
- NaturalHy (2004–2009) with a cost of €17 million for the study of hydrogen injection in the natural gas network
- HyMAT for the study of the influence of hydrogen on the properties of steels
- HySAFE (2004–2009) covering safety issues related to the use of hydrogen
- IdealHy to develop a concept to reduce the energy requirements for hydrogen liquefaction
- HELMET (2014–2017) for the study of high-temperature electrolysis (800 °C)
- H2SHIPS (2019–2022) is intended to demonstrate the feasibility of the use of hydrogen on vessels
- The US Department of Energy (DOE) funded, in 2021, 31 projects for hydrogen technologies in different sectors
- In 2022, the European project IPCEI Hy2Tech involving 35 companies in 15 countries has as an objective the decarbonisation of industrial processes and mobility. The 41 projects could be state subsidised by up to €5.4 billion and €8.8 billion from private investors.

7.2.3.2 INGRID project
This European project (high-capacity hydrogen-based green energy storage solutions for grid balancing), launched in 2012 and planned until 2016 with a cost of €24 million, evaluated the electricity produced by a wind farm in Troia, Italy. It included an electrolyser from Hydrogenics (1.2 MW to 240 Nm^3/h), 750 kg/39 MWh storage in hydrides (McPhy Energy), and a 60 kW fuel cell. In addition to experimenting with various parameters, simulation studies of various elements were carried out in order to optimise the operation of the installation (Figure 7.6).

The hydrogen produced was used for transportation (service station), injection into the natural gas network or industry.

What were the real benefits of these programmes?

The cumulative costs of European and national programmes amount to hundreds of millions of euros or US$. While some have provided results that can be exploited for experiments with operational P2G installations, others, often due to lack of information or clear conclusions, do not reveal whether they have really contributed to a better understanding of different phenomena or helped in the development of operational projects. Many countries or companies have similar projects but without international coordination. This leads to duplicate or projects that have been already completed in another country.

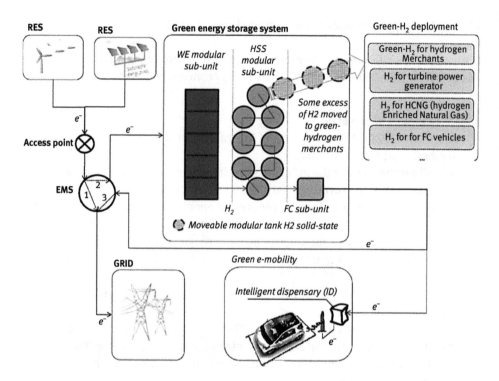

Figure 7.6: INGRID project diagram (ingridproject.eu).

7.3 Pilot projects

Many installations, mostly in Germany, are demonstrators of this technology, their power being not yet related to the future needs of storage of surplus renewable electricity. However, the operating conditions reflect those needed on a larger scale.

In 2022, existing installations can still be considered as pilot projects considering the size and total power of the electrolysers in service. Moving to industrial scale would require much powerful equipments (assuming the hydrogen will be produced from the surpluses for optimal energy efficiency). As an example, the European steel industry had a production of 152.6 million tonnes in 2021. Its decarbonisation would require 7.6 million tonnes of green hydrogen, assuming an hydrogen consumption of 50 kg/tonne of steel. According to the 2020 *European Commission, Joint Research Centre*, 6 GW of electrolysers would produce about 0.8 million tonne hydrogen per year. Thus, only the decarbonisation of European steel would require 57 GW of electrolyser power. In 2021, only a capacity of 135 MW was installed in Europe. The development of electrolysers is not correlated in terms of installed power with the production of renewable electricity (especially offshore wind with 28 GW in 2022 and 60–80 GW expected by 2030).

Many countries have presented a hydrogen roadmap or strategy for the next decades. However, few have moved to a high-power hydrogen infrastructure and often only prototype or small series units are planned or running.

7.3.1 Germany – the leader

Germany can be considered as a pioneer in exploring different approaches to P2G. In 2022, more than 30 facilities were operational and over 20 planned or under construction (Figure 7.7).

7.3.1.1 Enertrag (Prenzlau)

One of the first installations, built in 2009 and operational in 2011, combining many technological approaches, is that of the German company Enertrag in Prenzlau, near Berlin (Figures 7.8 and 7.9).

The installation (Table 7.2) offers a great deal of flexibility, allowing the choice of operating modes and covering different needs.

Operating modes available:
- maximum production of hydrogen;
- guaranteed electrical production with operation of cogeneration modules, if necessary, to stabilise electricity production;
- predictive mode based on 8 h weather forecasts to guarantee a given electricity production;
- regulation of the network with needs communicated 24 h in advance; and
- the injection of hydrogen into the natural gas network started at the end of 2014.

7.3.1.2 Thüga/Mainova (Frankfurt)

In 2012, the Thüga group of energy suppliers launched a P2G installation project. In 2013, the various modules were installed in Frankfurt-am-Main on a site of Mainova, the local energy supplier (Figure 7.10).

The installation includes:
- a 320 kW PEM-type electrolyser with a capacity of up to 60 Nm^3/h of hydrogen and
- a hydrogen injection unit in the natural gas network (between 2% and 5%) under a pressure of 3.5 bar.

Each module was installed in a 20 ft container, the electrolyser being supplied by the ITM Power (Figure 7.11).

In November 2013, hydrogen produced (130 kg/day) was first injected into the natural gas network under a pressure that did not require a compressor (Figure 7.12). The official inauguration took place in March 2014 and the project continued until 2016.

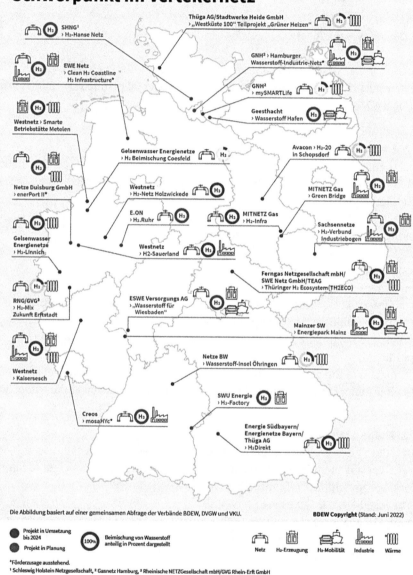

Figure 7.7: Power-to-gas projects planned or operational by 2024 in Germany as of June 2022 (BDEW).

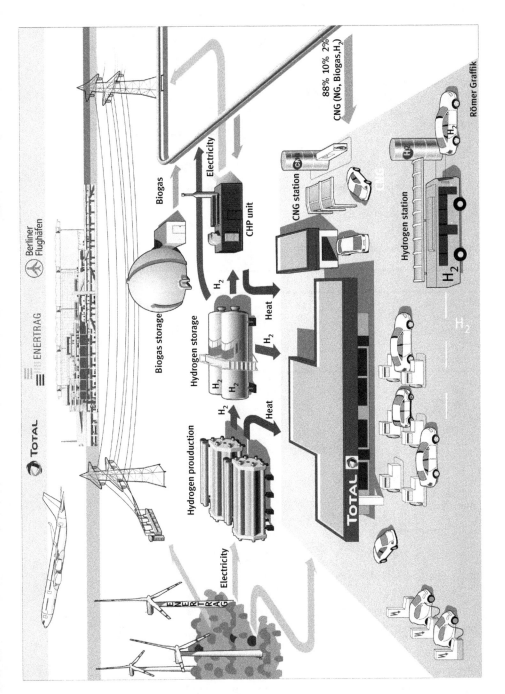

Figure 7.8: Power-to-gas Enertrag installation (Enertrag).

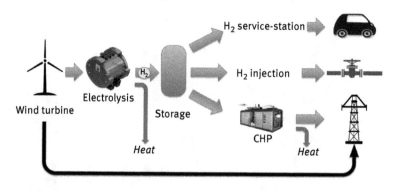

Figure 7.9: Schematic of the Enertrag installation.

Table 7.2: Enertrag installation components.

Equipment		
Wind turbine	3 × 2 MW	Enertrag
Alkaline electrolyser	500 kW to 129 Nm³/h	
H₂ compressor	31 bar	
H₂ storage	1,350 kg to 31 bar	
Combined heat and power unit	2 × 350 kW	

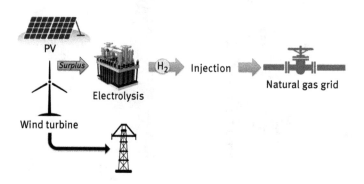

Figure 7.10: Schematic of the Thüga plant in Frankfurt.

In 2015, the overall hydrogen conversion efficiency was 77%. Monitoring and control were carried out from the Mainova control centre.

7.3.1.3 Viessmann: biological methanation (bioPower2Gas)

A biological methanation demonstration unit of MicrobEnergy, a subsidiary of Viessmann, was tested until the end of 2014 with a constant gas production (up to 5 Nm³/h) containing more than 98% methane (Figure 7.13).

Figure 7.11: A 320 kW electrolyser (Thüga/Mainova).

Figure 7.12: Hydrogen injection station in the natural gas network (Thüga/Mainova).

At the beginning of 2015, a unit installed at Allendorf at Viessmann's head office (Figures 7.14 and 7.15) used hydrogen produced locally by a 400 kW (1) PEM-type electrolyser supplied by Schmack Carbotech (also a Viessmann subsidiary). Hydrogen production capacity up to 220 Nm^3/h was fed to the biological methanation units (2) and (3) with a capacity of 55 Nm^3/h. The required CO_2 came from a biogas unit.

The next step allowing the methanation of 400 Nm^3 of hydrogen per hour was approved.

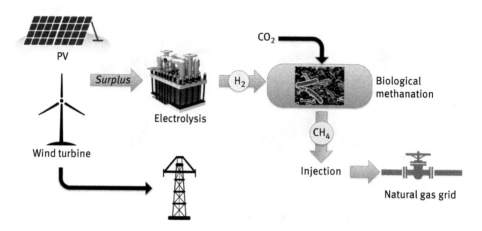

Figure 7.13: Concept of the Viessmann biological methanation plant.

Figure 7.14: Viessmann power-to-gas unit (Viessmann).

7.3.1.4 RH2-WKA

The RH2-WKA project (RH2-Werder/Kessin/Altentreptow), inaugurated in 2013, was designed to provide electricity primarily for the wind farm (Figure 7.16).

This project used the surplus electricity from 28 wind turbines (including 15 of 7.5 MW) with a total power of 140 MW supplied by Enercon. Three 1 MW alkaline

Figure 7.15: Biological methanation unit (Viessmann).

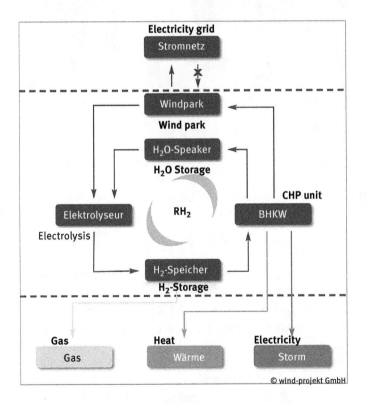

Figure 7.16: Schematic of the RH2-WKA installation (WIND Projekt GmbH).

electrolysers (Hydrogenics HySTAT®) can produce up to 210 Nm³/h of hydrogen. After drying and separating the oxygen, hydrogen is compressed to 300 bar (Hofer compressor) and stored in 120 steel bottles (810 kg, i.e. 9,500 Nm³ representing an energy of 27 MWh). Two cogeneration units (Senergie GmbH) of 160 and 90 kW (250 and 400 kW$_{th.}$ respectively) were modified to operate with hydrogen. The heat produced is used by a nearby farm (Figure 7.17).

Figure 7.17: RH2–WKA installation (WIND Projekt GmbH).

The hydrogen storage capacity allows the cogeneration units to operate at 28 h at maximum speed.

A second phase (RH2-PtG) involves the installation of a hydrogen injection unit into the natural gas network at a pressure of 25 bar.

7.3.1.5 Herten

Inaugurated in 2013 as part of the H2Herten project, this facility (Figure 7.18) used the hydrogen produced to supply electricity to a local minigrid.

As the wind farm is too far from the electrolyser, the electricity used is "virtual": when the wind farm generates excess electricity, the same quantity is taken from the grid to supply the electrolyser.

Electricity is partly stored in Li-ion batteries and partly used by the 280 kW (30 Nm³/h) electrolyser for an annual production of about 6,500 kg of hydrogen. This hydrogen is stored and used by a 50 kW fuel cell supplying electricity to a research centre and an industrial and commercial park (Mini Grid). This unit is designed to be able to operate autonomously, using the 250 MWh of electricity produced per year.

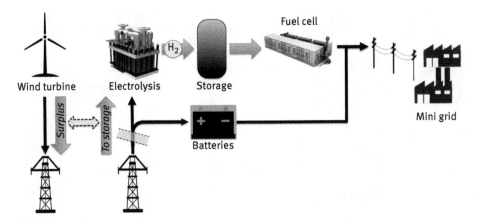

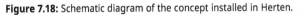

Figure 7.18: Schematic diagram of the concept installed in Herten.

The US fuel cell manufacturer Cummins has started the production of fuel cells (formerly Hydrogenics) close to Herten for hydrogen trains (Alstom Corada iLint). The yearly production capacity is 10 MW/year.

7.3.1.6 Mainz

The Energiepark Mainz (Figures 7.19 and 7.20), inaugurated in March 2015, is a research unit for the production of hydrogen from wind-generated electricity. It is located on the premises of the municipal energy provider (Stadtwerke) of Mainz (Figure 7.21).

Figure 7.19: Energiepark Mainz (Energiepark Mainz).

Figure 7.20: Power-to-gas installation of the Energiepark in Mainz (Energiepark Mainz).

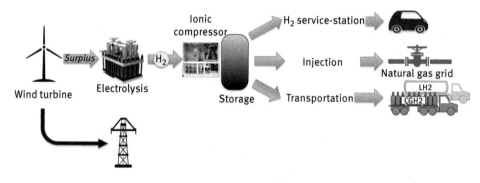

Figure 7.21: Concept of the Energiepark in Mainz.

It consists of a wind farm comprising four wind turbines with a total power of 10 MW, three electrolysers with 2 MW power each (Figure 7.22) supplied by Siemens (1,000 Nm³/h under a pressure of 35 bar), an ionic compressor from Linde and a storage in two 82 m³ tanks.

The hydrogen produced is either delivered by trailers or used by an on-site service station (Figure 7.23) or injected into the natural gas network (10% concentration).

In 2022, the main lessons learnt are: visible degradation of electrolyser cells, need of high-purity water and issues with seals in hydrogen circuit. The filling of trailers and the injection in the natural gas network are working perfectly.

Figure 7.22: Electrolysers (Energiepark Mainz).

Figure 7.23: Hydrogen service station (Energiepark Mainz).

7.3.1.7 Audi "e-gas" project

The "e-gas" concept developed by the car manufacturer Audi, based on the work of the ZSW Institute in Stuttgart, was commissioned in 2014. It consists of a methanation unit built by ETOGAS (bought in 2017 by the Swiss company Hitachi Zosen Inova), already involved in the experimental project ZSW. It is located on the site of the biogas unit of EWE Energie AG in Werlte, Lower Saxony, which provides CO_2 for methanation (Figure 7.24).

Electricity is produced by four wind turbines of 3.6 MW each. The three alkaline electrolysers with a total power of 6 MW operate intermittently in case of excess electricity and can produce up to 1,300 Nm^3/h of hydrogen for a water consumption of 1.3 m^3. The heat from the electrolysers is recovered for the biogas unit (Figure 7.25).

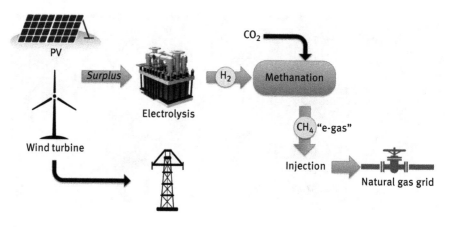

Figure 7.24: Audi e-gas concept.

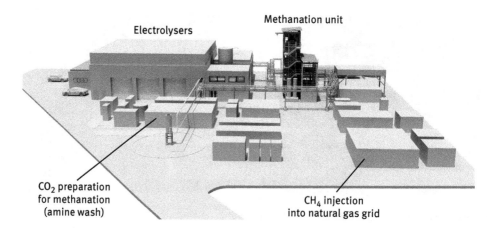

Figure 7.25: Audi power-to-gas unit (Audi).

After purification, compression and intermediate storage, hydrogen is used by the methanation unit (Figure 7.26), with a production capacity of 300 Nm³/h and a methane concentration of more than 99%. The heat produced converts the cooling water into steam (250–300 °C under 70 bar). The methanation unit is operational in less than 5 min.

Beginning 2021, the unit has been transferred to the Kiwi AG which provides the synthetic fuel to the Alternoil company which distributes it through 40 LNG refuelling stations. The container vessel ElbBLUE (152 m long and with a capacity of 1,000 20 ft containers) also uses the synthetic fuel as 50/50 mixture with LNG. However, the e-fuel still costs about 5 times more than LNG.

Originally, a service station has been planned to deliver the methane produced directly on site. However, Audi has focused on the use of hybrid vehicles Audi A3 or A4 g-tron bi-fuel running on gasoline or natural gas available at the 600 NGV service

Figure 7.26: Methanation unit (Audi).

stations in Germany. A specific card system allowed payment, counting the volume of gas used. CO_2 emissions from these vehicles are then calculated on the basis of natural gas consumption by the vehicle and production by the P2G unit. The result is an average corrected emission of 20 g/km (well to wheel analysis) per vehicle, and the annual methane production corresponding to the consumption of 1,500 vehicles travelling each 15,000 km (Figure 7.27).

7.3.1.8 Oberhausen
Air Liquide/Siemens Energy is installing a 20 MW PEM electrolyser (upgradable up to 30 MW) to produce 2,900 tonnes of hydrogen per year in a first step starting from 2023. This hydrogen will be used for the North-Rhine Westphalia chemical industry and decarbonising refineries or steel industry (Tyssenkrupp plant in Duisburg). The cost of the project is estimated to be €30 million.

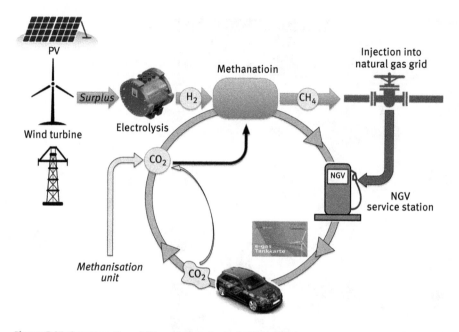

Figure 7.27: Compensation of CO_2 emissions by Audi NGV vehicles.

7.3.1.9 WindH2 and GrinHy2.0 (Green Industrial Hydrogen)

The WindH2 is part of the SALCOS (SAlzgitter Low CO_2 Steel making) project. The objective of SALCOS is to reduce the CO_2 emissions by 95% by 2033. Seven wind turbines (total power of 30 MW) were installed in 2020 and feed two Siemens PEM electrolysers (2×1.25 MW, 450 m³ hydrogen per hour).

For the GrinHy2.0, started in 2019, a high-temperature electrolyser (SOEC from Sunfire running at 850 °C and with steam supplied by the steel production process) is used to produce hydrogen for the annealing process (heat treatment to change physical and/or chemical properties of an element). The 720 kW electrolyser will produce up to 200 Nm³ hydrogen per hour. The ramp-up time is less than 5 min.

In the first part of the project (GrinHy 2016–2019), the prototype produced 100 tonnes of green hydrogen after 13,000 working hours. The overall efficiency was 84% (LHV).

7.3.1.10 Energiepark Bad Lauchstädt

The €140 million project (HYPOS – Hydrogen Power Storage & Solutions East Germany) concerns a 30 MW electrolyser capacity in 2024 with the electricity being supplied by a 50 MW wind park (Figure 7.28). The hydrogen will be delivered to Leuna through a 20 km natural gas pipeline. A storage in a cavern (50 million m³) will be targeted for 2026. Sunfire will supply the alkaline electrolyser.

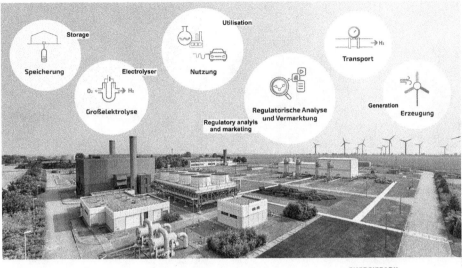

Figure 7.28: Energeipark Bad Lauchstädt (English legend from the author).

7.3.1.11 Wunsiedel WUN H2
In 2022, the Bavarian city of Wunsiedel producing at least 100% of its electricity (solar, wind and biomass) has started an 8.75 MW Siemens Silyzer 300 electrolyser (capacity of 1,350 tonne/year) to use also the excess electricity. Hydrogen will be supplied to local customers. A lithium battery of 10 MWh (8.4 MW power) is operational since 2018.

7.3.1.12 "GreenRoot"
The Dutch company HyCC and the German gas supplier VNG AG have signed an agreement in 2022 to develop hydrogen production in an industrial scale in Central Germany. The high concentration of chemical plants around the city of Leuna and the decarbonisation objectives are important factors for this project.

7.3.1.13 E.ON – Falkenhagen
E.ON, through its subsidiary E.ON Gas Storage in partnership with Swissgas, inaugurated an installation in Falkenhagen ("WindGas" project) near Hamburg in 2013 (Figures 7.29–7.31).

Figure 7.29: Falkenhagen power-to-gas unit (E.ON).

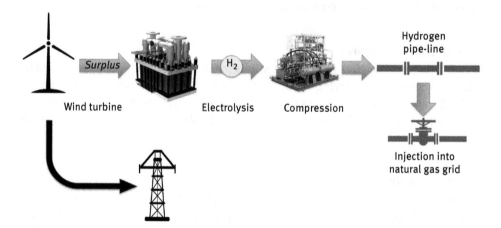

Figure 7.30: Concept of the Falkenhagen power-to-gas unit.

It uses the excess electricity from a 400 MW wind farm and consists of:

- a series of six electrolysers from Hydrogenics with a power of 2 MW_e producing up to 360 Nm^3/h of hydrogen under 10 bar by consuming 288 L/h of water,
- a hydrogen compressor (55 bar),
- a 1.6 km pipeline between the P2G unit and the natural gas injection and
- an injection unit into the natural gas network (up to 2% hydrogen under 55 bar)

Figure 7.31: Electrolysers of the Falkenhagen power-to-gas unit (Hydrogenics Corp.).

The yield (electricity used/injection in the natural gas network) was 58% in 2015 and should be improved by exploiting the oxygen that is released into the atmosphere and the heat produced (e.g. adding one unit of methanation) and an increase in the concentration of hydrogen in natural gas.

Another E.ON project, launched in 2014, still under the "WindGas" model, is located in Hamburg/Reitbrook and consists of a 1 MW$_e$ PEM-type electrolyser supplied by Hydrogenics and producing 265 Nm^3/h of hydrogen which is injected into the natural gas network.

7.3.1.14 Compact P2G unit

The German company Exytron has developed a compact P2G unit. A demonstrator (21 kW to 4 Nm^3/h electrolyser) was commissioned in Rostock in 2015. Field testing began in the late 2016 in the town of Alzey near Mannheim for a 37-housing residential unit (Figure 7.32).

The complete system (Figure 7.33) contains the alkaline electrolyser (62 kW to 10 Nm^3/h) using a photovoltaic park of 125 kW, a methanation unit (2.5 Nm^3/h), condensing boilers, combined heat and power (CHP) unit, hot water storage and control unit.

The houses will also be supplied with heat from district heating. "Green" electricity is provided by the grid when needed.

In 2019, a unit was installed in Augsburg, Germany, to provide electricity and heat to 70 flats. Exytron is building in Lübesse, Mecklenburg–West Pomerania, an "energy factory" with a 4 MW electrolyser. The unit will produce hydrogen that will also be converted into methane (up to 1200 tonne/year) or e-fuel. Two CHP units will generate electricity, and heat will be used by the local district heating.

Figure 7.32: Compact P2G unit (Exytron).

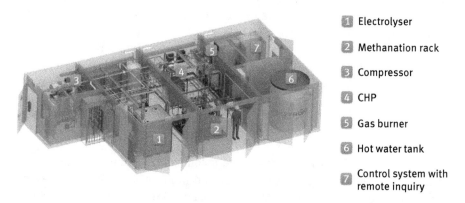

1 Electrolyser

2 Methanation rack

3 Compressor

4 CHP

5 Gas burner

6 Hot water tank

7 Control system with remote inquiry

Figure 7.33: Complete system for the dwelling unit.

7.3.1.15 Hydrogen separation from blend with natural gas

In 2022, the Linde Engineering and Evonik companies started running in Dormagen the first production unit for recovery of hydrogen from blends containing between 5% and 60% hydrogen. First, separation (HISELECT® membrane) gives an hydrogen purity of 90%. The following step using pressure swing adsorption (PSA) brings the purity to 99.9999%.

7.3.1.16 "Gigafactories"

Numerous projects involving the production of hydrogen at a large scale are planned.

The energy supplier RWE is targeting with Linde at least 2 GW of electrolyser capacity by 2030. In the frame of the GET H2 project, a 300 MW electrolyser capacity will be installed by 2026 in **Lingen**. The first step (2024–2025) will concern two PEM electrolysers of 100 MW power each. Electricity will be supplied by the offshore parks in the North Sea.

The chemical company INEOS and the industry park manager Curenta (project ChemCH$_2$ange) are planning a 100 MW electrolyser capacity near **Cologne**. Hydrogen will be used by the INEOS ammonia and methanol production plant.

In **Lubmin** (Mecklenburg-Western Pomerania), the German HH2E AG and the Swiss MET Group are planning a 100 MW production unit for 2025 (50 MW electrolyser and 200 MWh battery capacity to increase the working time of the electrolyser) with a capacity of 6,000 tonne/year with the objective to reach 60,000 tonnes in 2030 with a 1 GW power.

7.3.2 France

Paradoxically, while France still relies on nuclear power (which accounts for about 78% of electricity production in 2017), some innovative P2G projects that have been carried out are under way or planned.

7.3.2.1 MYRTE project

It is a research and feasibility project located in Corsica, in Vignola near Ajaccio (MYRTE – Mission hYdrogène Renouvelable pour l'inTégration au réseau Electrique – *Renewable Hydrogen Mission for Integration into the Electrical Network*) with the objective to study the stabilisation of the electrical network (Figures 7.34 and 7.35).

At a cost of €24 million, the programme was launched in 2006 and led by the University of Corsica, the HELION company and the Commissariat à l'énergie atomique – Commission of nuclear energy. It was inaugurated in early 2012 (Figure 7.36), which consists of the following elements:

– a photovoltaic park of 550 kW,
– a 200 kW electrolyser supplying hydrogen (40 Nm^3/h) at a pressure of 35 bar,
– storage units for hydrogen (four tanks) and oxygen (two tanks) under 35 bar, each tank having a capacity of 28 m^3,
– heat storage and
– a PEMFC fuel cell of 200 kW.

The aim of the project is to define strategies for the management and stabilisation of the electricity network (using the ORIENTE software – Optimization of Renewable Intermittent Energies with Hydrogen for Autonomous Electrification). The option studied is

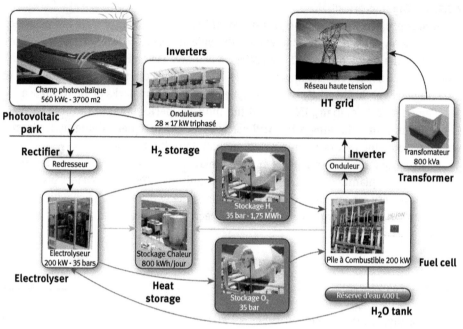

Figure 7.34: Project concept MYRTE (University of Corsica).

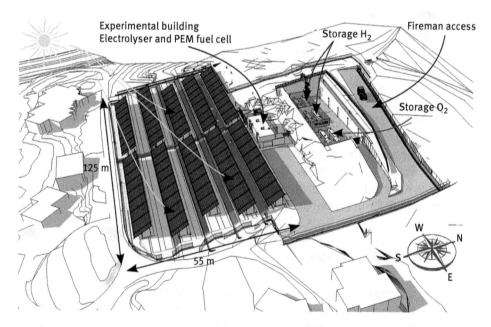

Figure 7.35: Schematic of the MYRTE project installation (University of Corsica).

Figure 7.36: MYRTE installation (University of Corsica).

to smooth wind power production; another is to use the fuel cell to generate extra electricity in case of peak demand. The University of Corsica operates the facility and coordinates R&D activities.

In 2022, the pilot will be extended by Li-ion batteries and improved the fuel cell. It will also provide cooling (using an adsorption unit) and hydrogen for a refuelling station.

7.3.2.2 GRHYD project

The GRHYD Demonstrator (Gestion des Réseaux par l'injection d'HYdrogène pour Décarboner les énergies – *Network management through hydrogen injection for decarbonising energy*) coordinated by the energy provider Engie was launched at the end of 2013 and has two objectives:

- injection of hydrogen produced by electrolysis into the local natural gas network and
- evaluation of hythane fuel, a mixture of hydrogen (6–20%) and natural gas, for transport.

It is implemented in the Dunkerque community, north of France. The required hydrogen is produced by electrolysis (50 kW PEM-type electrolyser), compressed and stored in metal hydrides.

The first part concerns a district of 200 low-energy dwellings delivered at the end of 2015 and an hospital which is fuelled in 2018 by a mixture of natural gas and hydrogen (up to 20%) to cover heating, hot water and cooking needs. No equipment has been modified. In 2022, the feasibility has been proven (distribution network and available equipments, lower emissions).

7.3.2.3 Jupiter 1000 project

French gas grid operator GRTgaz coordinates the demonstration project (Figure 7.37) for methane production from surplus electricity, generated by four wind turbines (10 MW) and located on the industrial zone of Fos-sur-Mer (INNOVEX platform). The two evaluated electrolysers supplied by McPhy Energy (PEM and alkaline of 500 kW each) will produce up to 200 Nm^3/day of hydrogen which will be used partly for direct injection in the natural gas grid, and partly for methanation (25 Nm^3/day) whose equipment is supplied by the Atmostat company. CO_2 is captured from a close metallurgical plant. The methane produced will also be injected into the gas network. This project was operational in 2020. In 2022, the maximum concentration of hydrogen injected was 6% but usually in the range of 1–2%. It took 8 years and €23 million to reach this first step.

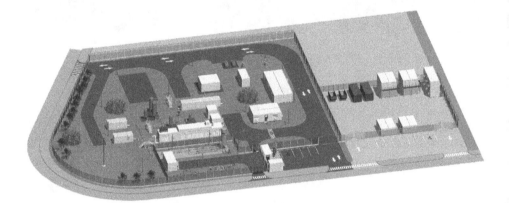

Figure 7.37: Layout of Jupiter 1000 power-to-gas installation.

7.3.2.4 Industrial hydrogen production units ("Gigafactories")

The H2V and Lhyfe (created in 2017) companies are moving towards building large green hydrogen production units. The H2VN project in Normandy, overtaken in 2021 by Air Liquide, planned for 2025, will use "certified" green electricity. The 200 MW of Siemens Energy PEM electrolysers will produce up to 28,000 tonne/year for industrial use. Another H2V project in Thionville (east of France), planned to be operational in 2026, will have four electrolyser units with 100 MW each for an annual production of 56,000 tonnes of hydrogen. H2V is considering other locations in France. The Lhyfe projects also involve large-scale hydrogen production. A project with Lhyfe as an integrator in Skyve, Denmark, will have up to 100 MW of electrolysers (6 MW end of 2022), fed by green electricity.

Offshore hydrogen production is also explored. A Lhyfe experimentation 20 km from the French coast with a floating unit demonstrator Sealhyfe has taken place in 2022. Electricity was supplied by a prototype floating wind turbine. The electrolyser supplied by Plug Power had a capacity of 400 kg hydrogen/day. At another scale, Tractebel, subsidiary of the French company Engie, presented a concept around a platform for the 400 MW electrolysers in the North Sea. The wind park would supply either electricity or hydrogen, when there is an electricity surplus, through a submarine pipeline. The windpark production will thus be optimised.

7.3.3 Other countries

7.3.3.1 Denmark

As early as 2006, Denmark began studying the production of hydrogen from renewable electricity. The first operational system in 2007 consisted of two 4 kW PEM and two PEMFCs with 2 and 7.5 kW power. The oxygen and hydrogen produced were stored separately in two tanks (25 and 12.5 m^3) under low pressure. Hydrogen was used by fuel cells and oxygen by the wastewater treatment plant.

Experimentation at the level of a district began in 2007 in the locality of Vesten-skov on the island of Lolland. An electrolyser used the electricity of a wind turbine, and the hydrogen produced, after storage, was distributed through a specific network initially feeding 5 houses and then extended to 30 others. Low-power fuel cells in-stalled in housings (PEMFC from the Danish company IRD) used directly pure hydro-gen to produce electricity (0.9–2.0 kW) and heat (0.8–2.0 kW).

The "**BioCat**" demonstrator project (P2G via Biological Catalysis), based on the transformation of hydrogen into methane using microorganisms in the presence of CO_2, has been launched in February 2013, and since April 2016 the 1 MW installation in Avedøre is operational, supplying methane with a purity of 98%. Pilot plants started in 2019 in Colorado, USA, and Solothurn, Switzerland, with grid injection. The next step is a 50 MW$_e$ unit.

The HyBalance project (Figure 7.38), European and Danish funded, inaugurated in 2018 in Hobro, north of Denmark, used a 1.2 MW PEM electrolyser to produce hydrogen for an industrial complex or transportation (fleet of fuel cell taxis in Copenhagen). The other objective was to test grid balancing where the electrolyser showed a ramp up or ramp down of less than 10 s. The project was closed in 2020 and had a total cost of €15.8 million.

In an industrial park (GreenLab) in Skive, a 6 MW alkaline electrolyser unit from the Danish Green Hydrogen Systems will be the first step for a project that has a 100 MW target by 2024.

Figure 7.38: HyBalance installation (HyBalance).

In Brande, west of Denmark, a 430 kW electrolyser from Green Hydrogen Systems fed by a 3 MW wind turbine from Siemens Gamesa will produce hydrogen onsite. The wind turbine is not connected to the electrical network but only used by the electrolyser. A 560 kWh Li-ion battery acts as a buffer to store electricity not used by the electrolyser and allows operation when there is no or not enough wind. The production started in 2021.

An optimised approach is needed for off-grid installations

However, the Brande project not being connected to the electrical network shows the limits of such option: assuming that the wind turbine will run for example at 75% of its power (2 MW_e) for 3 h producing an energy of 6 MWh. Simultaneously, the battery will be loaded within 3 h (assuming a load time of 3 h) requiring 560 kWh (186 kW during 3 h) and the electrolyser running at 100% capacity will require 430 kW. The global outcome is the power "loss" of 1,384 kW (2000-186-430) as the installation cannot export the electricity surplus. In this case, the wind turbine is over-dimensioned.

The "Green Fuels for Denmark (GFDK)" project will bring in a first step 100 MW of electrolysis in 2025 with carbon capture and storage (CCS). The target is to produce 50,000 tonnes of e-fuels for transportation. The first phase planned for 2023 will see a 10 MW electrolyser used to produce hydrogen for transportation. The second phase (2023–2027) will increase to 250 MW of electrolysers to produce e-methanol and e-kerosene.

7.3.3.2 The Netherlands

A demonstrator has been tested in Rozenburg (part of the Rotterdam agglomeration) until 2015. It included an electrolyser and a methanation unit housed in two containers and a CO_2 storage unit for methanation.

The electricity was supplied by photovoltaic panels and by the grid, if necessary. The installation included:

– a PEM-type electrolyser with a production capacity of 1.0 Nm^3/h of hydrogen (2.27 kg/day) under a pressure of 13.8 bar, with a yield of about 47%;
– a drying unit (PSA);
– a methanation unit with four reactors with an overall efficiency of 73% (Figure 7.39) and
– a chromatograph to measure the composition of gas from the methanation unit and verify that it meets the standards for injection into the natural gas network.

Figure 7.39: Demonstrator of Rozenburg (DNV GL).

Taking into account the initial checks, the electrolyser is operational after 4 min (hydrogen production). The methanation unit takes approximately 37 min before methane is supplied (Figures 7.40 and 7.41).

Figure 7.40: Methanation units (DNV GL).

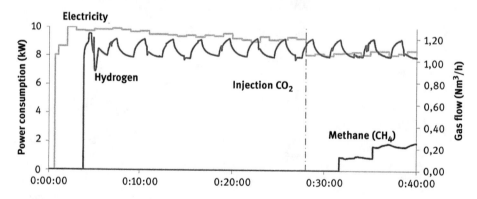

Figure 7.41: Starting time of various processes (DNV GL).

During this experiment, methane had to meet the following criteria:
- Wobbe index between 44.10 and 44.41 MJ/m^3;
- less than 0.5 mol% oxygen;
- less than 10.3 mol% CO and
- less than 0.1 mol% hydrogen.

A series of sensors checked the parameters and the injection could be stopped as soon as one of them did not meet the specified values. This experiment demonstrated the viability of the concept. Due to the small size of the plant, the yields (especially of the electrolyser) could be improved.

The other **Hystock** project was initiated in 2017 by Gasunie New Energy and Gasunie EnergyStock with a pilot unit consisting of a 1 MW electrolyser. The facility used electricity from a photovoltaic plant with 5,000 modules and was located near the underground natural gas storage site in Zuidwending (Groningen province). The hydrogen produced was compressed and stored in cylinders for use in transportation or industry. In 2022, a preliminary study for hydrogen storage in the existing natural gas storage will be evaluated, and if results are positive, hydrogen storage will start in 2026. Gasunie plans to convert by 2027 part of natural gas network to carry hydrogen using up to 85% of existing pipes.

In 2022, the port of Rotterdam launched an ambitious project involving production or import of hydrogen at a large scale (Figure 7.42). The 200 MW of electrolysers by the Shell company will produce up to 60 tonnes of hydrogen per day. The H2-Fifty project involves bp with 250 MW of electrolysers to be used by refineries and industries. These are the first steps. An hydrogen pipeline network will link Rotterdam to the rest of the Netherlands and Germany.

7.3.3.3 The UK

In 2020, the *Ten Point Plan* projected 5 GW of electrolysers for low carbon hydrogen by 2030. In 2021, the *UK Hydrogen Strategy* and a concrete roadmap 2030/2050 confirmed and extended the 2020 plan by including hydrogen network and storage. In 2016, a project was started in Levenmouth (wind turbine of 750 kW, alkaline electrolyser of 30 kW, storage of 11 kg of hydrogen) with hydrogen used by a service station (vehicles with range extender) and for production of electricity and heat (10 kW fuel cell and hydrogen-fired boiler). An extension is the *H100 Fife* project that will see in 2023 hydrogen for heating and cooking for 300 homes. Hydrogen will be produced by the green electricity from a 7 MW wind turbine feeding an electrolyser. Hydrogen will be stored in six tanks and supplied through a dedicated network to the selected homes.

The BIG H_2IT in the Orkney Island will involve producing green hydrogen (50 tonne/year) on the Eday (0.5 MW PEM electrolyser) and Shapinsay (1.0 MW PEM electrolyser) islands, storing it on tube trailers and transporting it to the mainland Orkney. In Kirkwall, a 75 kW fuel cell will supply heat and power. Hydrogen will also be used for a refuelling station.

As part of the **HyDeploy** project launched in 2017 and funded by Ofgem (Office of Gas and Electricity Markets) and coordinated by National Grid, ITM Power will provide a 500 kW electrolyser to produce hydrogen that will be injected into the gas grid of the Keele University campus with nearly 340 residential, research or industrial buildings. At the end of 2021, the project was operational in 100 homes and 30 university buildings without any change to piping and equipments.

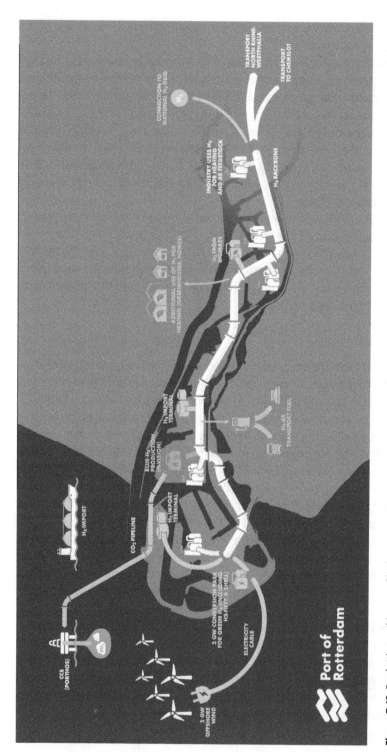

Figure 7.42: Production and import of hydrogen at the Rotterdam harbour (Port of Rotterdam).

Other UK projects are:
- Dolphy Project – offshore hydrogen production using seawater,
- Tees Green Hydrogen near Teeside offshore wind farm and a new solar farm and
- Gigastack with electricity from the Ørstedt offshore wind park supplying an ITM electrolyser producing hydrogen for a refinery.

7.3.3.4 The USA

The WIN2H2 project (2007–2010), managed by the NREL and Xcel Energy, was the first step towards the production of electricity and the use of hydrogen for transportation.

A P2G project (SoCalGas – Southern California Gas Company, NREL and Electrochaea) located in Golden, Colorado, has been scheduled from September 2014 to March 2016 in the NREL laboratories with a 150 kW electrolyser to produce hydrogen. CO_2 was transformed into methane through biological methanation. The gas produced was used in a fuel cell.

SoCalGas launched a bioreactor demonstrator in early 2017, supported by the DOE. It was installed at the NREL Energy facility (Energy Systems Integration Facility). The design, fabrication and validation of a bioreactor was commissioned in 2019. In 2023, the SoCalGas' biomethanation reactor will be transferred to Clinton, Maine.

The Advanced Clean Energy Storage Project (ACES) project in Utah, USA, with Mitsubishi Power Americas, Inc., as lead sponsor will use alkaline electrolysers with a total power of 220 MW to produce green hydrogen. It will be stored in two salt caverns with a capacity of up to 5,500 tonnes and used by a 840 MW gas turbine (CCGT). It will first run on a blend of 30% hydrogen starting in 2025 and increase to 100% by 2045.

In 2022, the DOE Hydrogen Program considered hydrogen production from all possible sources (natural gas, coal, nuclear or renewables). The *Bipartisan Infrastructure Law* (BIL)) signed in 2021 funded US $9.5 billion over 5 years for clean hydrogen (not necessarily green) programs.

7.3.3.5 Canada

In 2015, Glencore inaugurated a Mini Grid project for the Raglan nickel mine supplying the 25 kV local power grid. An alkaline electrolyser of 315 kW uses the electricity of a 3 MW wind turbine (the fluctuations are filtered by a flywheel) and supplies hydrogen stored under pressure in three tanks with a PEM fuel cell of 200 kW which supplement the battery storage (250 kW/250 kWh) or the diesel generator.

Enbridge Gas Distribution in partnership with Hydrogenics Corp. has conducted a project with a 2 MW electrolyser in Markham, Ontario, which was operational in 2018. In 2022, hydrogen blend (up to 2% only) supply to 3,600 customers (homes and business) in Markham was operational.

7.3.3.6 Japan

The first P2G project of the New Energy and Industrial Technology Development Organization (NEDO), which began in 2015, was located in Kofu City, Yamanashi Prefecture, west of Tokyo. The electricity was produced by a 1 MW photovoltaic park, solar modules on the roof of the visitor centre and a hydropower generator. The excess electricity was used by an electrolyser from Kobelco Eco-Solutions Co. to produce hydrogen stored under pressure which powers a Panasonic Fuel Cell that compensates for the lack of electricity in the event of low solar production.

In 2020, a 10 MW electrolyser was operational in the town of Namie in the frame of the Fukushima Hydrogen Energy Research Field (FH2R). Electricity was supplied by a 20 MW_{peak} PV farm. Production was up to 1,200 Nm^3/h.

However, Japan is investing more in blue (from natural gas) or black (from coal) hydrogen imported from the United Arab Emirates, Brunei (as LOHC) or Australia (as LH2).

7.3.3.7 Switzerland

The Aarmat hybrid plant (Regio Energie Solothurn) operational since 2015 combined conventional units (natural gas-fired boiler and CHP, heat storage) with a P2G installation. A PEM electrolyser (350 kW) supplied by Proton Onsite produced up to 60 Nm^3/h hydrogen stored under 30 bar in tanks (180 m^3) and then injected into the natural gas grid. In the frame of the European projects "Horizon 2020" and "Store & Go," a biological methanation unit was planned in 2018. In 2019, the unit supplied by Electrochaea injected the produced methane in the natural gas network.

The German company Hitachi Zosen Inova Schmack GmbH (HZI Schmack) has inaugurated the first industrial biological methanation unit in 2022 in Dietikon (Figure 7.43). CO_2 is captured from sewage gas.

Figure 7.43: Biological methanation unit with the reactor installation on the left picture (Hitachi Zosen Innova AG).

The energy supplier GAZNAT experimented in 2020 a 20 kW one-stage methanation (CO_2 is supplied by an industrial site) unit using hydrogen from an electrolyser. The methane (79,000 kWh/year) is then injected in the natural gas network. GAZNAT is planning a 200 kW methanation unit.

7.3.3.8 Chile

Even if green hydrogen production is limited (INDURA Lirquèn through electrolysis since 1996), Chile is seen especially by Germany as a potential source of green hydrogen (estimation of 160 million tonne/year) or e-fuels: the Haru Oni project, led by Enel Green Power Chile, will use electricity from a 3.4 MW wind turbine to power a 1.25 MW electrolyser. Synthetic fuel will be produced and exported for the car manufacturer Porsche. Chilean companies Engie Latam SA and Enaex SA are planning 26 MW of electrolysers (3,200 tonne/year of hydrogen) associated with a green ammonia plant by 2025.

7.3.3.9 China

Chinas' hydrogen production in 2022 was about 33 million tonnes (3% from electrolysis) representing one third of the world production. It has, however, ambitious plans for green hydrogen: 30 million tonnes have been planned for 2040 and 60 million for 2050 although targets for the short-term (2025) are still very low. To reach these goals, important investments are planned in the frame of the 2022 *Medium- and Long-Term Plan for Hydrogen Energy Industry Development (2021–2035)* in R&D, projects developments and industrial structure.

In 2022, about 1 GW of electrolysers are operational and 38 GW are planned for 2030. Different clusters involving electrolyser companies, universities and research institutes have been created (Dalian or Handan for example). Local governments (Hebei, Sichuan, Ningxia Hui, Inner Mongolia etc.) have also their own programs.

Electrolyser manufacturer industry is active representing about 35% of world manufacturing capacity: PERIC Hydrogen Technologies (alkaline and PEM), Cockerill Jingli Hydrogen (alkaline) or Tianjin Mainland Hydrogen Equipment (alkaline).

In 2022, there are about 120 green hydrogen projects financed mainly by energy operators like Sinopec. Most of them are of small power but some are the most powerful worldwide. Ningxia Baofeng Energy Group's 150 MW alkaline electrolysers operational in 2021 are powered by a 200 MW solar farm. Sinopec is building a 260 MW alkaline electrolyser plant fed by a 300 MW PV farm that should operate in 2023 in Xinjiang. Hydrogen will be delivered by a pipeline to an oil refinery.

With its important wind and solar potential, the production of green hydrogen could fulfil the governmental objectives. However, the size of the country and the availability of water could limit production in some provinces. The Belt and Road Initiative could offer opportunities either to export hydrogen or to import it if production costs are lower than in China.

7.3.3.10 European Union

Independently of national programmes or projects, the EU is also subsidising other projects. The IPCEI Hy2Use with a public funding of €5.2 billion involves 13 countries. The objective of the 29 companies is to develop hydrogen projects like electrolyser, infrastructure, integration in industrial processes etc. Timeline is project related: large-scale electrolyser will be operational in 2026, innovative technologies in 2027 and completion of the 35 defined projects in 2036.

7.3.3.11 Other countries

If many countries have already specified or voted hydrogen strategies, the projects planned for short- or mid-term are not in line with the needs to decarbonise the different energy sectors. Even if the governmental investments seem high (billion euros or US$), once divided by the duration and the number of programs, they are not sufficient to initiate a drastic increase of hydrogen production: the German government, for example, has planned €8 billion but split between 62 industrial projects.

Some countries, like South Korea, have a large park of fuel cell for electricity production and are pushing hydrogen for transportation. However, P2G is still at the beginning, the country's policy being not clear as to whether hydrogen (green or grey) should be produced or imported.

Hydrogen "Gigafactories"

The "Gigafactory" concept (in fact hydrogen production units of tens or hundreds of MW) is spreading as hydrogen need is increasing. The Holland Hydrogen I from the Shell company is planned for 2025 with a 200 MW electrolysers capacity in the port of Rotterdam. The daily production will be up to 60 tonnes. The Danish project HySynergy planned in Fredericia started with electrolysis power of 20 MW in 2022, moving to 300 MW in 2025 and 1 GW in 2030. Always in Denmark, the H2RES project near Copenhagen is targeting an electrolysis capacity of 1.3 GW in 2030.

7.4 Comparison of current projects

7.4.1 Technologies

The main sources of renewable electricity are either predominant wind, solar or a combination of both.

The alkaline electrolysers remain used although their efficiency is lower than the one of the PEM type. The advantage is lower cost and proven technology. AEM technology is however finding increased development.

7.4.2 Applications

Considering the projects and demonstrators in Germany which are more representative, the applications can be divided into four categories (Table 7.3):
–   Injection of hydrogen into the natural gas network
–   Direct use for service stations
–   Methanation (thermochemical or biological) and injection of methane into the natural gas network
–   Conversion of hydrogen (or methane) into chemical compound or e-fuel (power-to-liquid)

Table 7.3: Comparison of power-to-gas first-generation and 2022 projects.

Country	Project	Electrolyser	Power MW	Status
Germany	*Enertrag-Prenzlau*	*Alkaline*	*0.5*	*Operational 2011*
Germany	*Energiepark Mainz*	*PEM*	*6*	*Operational 2015*
Germany	*Thüga/Mainova*	*PEM*	*0.4*	*Operational 2013*
France	*Myrte*	*PEM*	*0.2*	*Operational 2012*
France	*Jupiter 1000*	*Alkaline + PEM/McPhy*	*2x0.5*	*Operational 2020*
France	H2VN Normandy	Alkaline/Hydrogen pro	2x100	2025
Norway	Yara	PEM/Sunfire	24	2023
Finland	P2Xsolutions	Alkaline/Sunfire	20	2024
Germany	Lubmin	NA	50	2025
Germany	Oberhausen	PEM/Siemens Energy	20	2023
Denmark	Skive	Alkaline/Green Hydrogen	24	2023

This comparison shows the change of scale of electrolyser power in 2022 for the new projects.

7.5 Experimental results

The on-going experiments that serve primarily to demonstrate the feasibility of this technology can only be considered currently as prototypes in terms of costs.

The key points of the installations for their exploitation are the commercialisation of hydrogen or methane produced or the (re)conversion of these gases into electricity.

The first experiments and results associated with P2G show the following average yields for each equipment:
–   alkaline electrolysis: 70–75%,
–   PEM electrolysis: ≥80%,
–   SOEC electrolyser: ≥84%,
–   methanation: ≥80%,

- methanation and heat recovery: ≥90%,
- electricity production (CHP or fuel cell): 35–50% and
- electricity production (CHP or fuel cell) and heat recovery: 80–85%.

These results show the technical feasibility from the point of view of efficiency, especially if the heat produced during the different stages is valued (district heating, biogas plants, farms, greenhouses or industry).

8 Financial approach to power-to-gas

8.1 Hydrogen conversion capacity versus expected surpluses of renewable electricity

The rapid expected growth of renewable electricity (sun and wind) by 2030 and even more by 2050 will lead to generation surpluses. What to do if those surpluses cannot be used: either insufficient national transport capacity or no possible export (neighbouring countries having also a surplus or limited exchange capacity)?

In Germany, for example, offshore wind parks are located mainly in the North Sea, whereas southern regions (Bavaria or Baden-Württemberg for example) are large consumers with low renewable generation. The construction of high-tension power lines is lagging behind schedule due to protests and lack of investment those last years. Over 7,500 km of transmission grid lines will need to be upgraded or newly built. From the 1,655 km adopted in 2009, 767 km were in operation in 2019; and from the 2,494 km adopted in 2013, 183 km were in operation in 2019.

Given the electricity surpluses expected, their storage and eventually their conversion to electricity will require equipment (especially electrolysers) of high power. It is in these areas that progress is expected in order to store the maximum of the surplus.

8.1.1 Capacity of electrolysers

In 2022, the maximum power of PEM electrolysers reaches 20 MW. For the electricity surpluses expected over the next few decades (Table 8.1), the power of the installed electrolysers should theoretically be able to cover the surplus at peak generation.

Table 8.1: Potential surplus of electricity from renewable sources.

Country		Yearly estimated surplus	
Germany	Excess in 2012	400 GWh	
	2020 scenario	25 TWh	
	2050 scenario	162 TWh	85% renewable
Denmark	2020 scenario	3 GWh	Over 350–450 h
USA	2050 scenario	110 TWh	Production capacity 1,450 GW

A 20 MW electrolyser can only convert surplus electricity from three 6 MW wind turbines operating at maximum power. The compromise to be found will be between the maximum power of the electrolysers and these peaks of surplus: recovering all excess electricity will lead to a relatively low utilisation of the maximum capacity of the

https://doi.org/10.1515/9783110781892-009

electrolysers. However, a too low electrolyser power will not allow the "recovery" of the potentially possible maximum of electricity. One option would be a regional cluster of smaller electrolysers where the surplus of electricity can be directed.

It is possible to consider units of higher power or a combination of less powerful units. In any case, to maximise the conversion of all the surplus electricity, the maximum power will have to be able to absorb the peaks of surpluses which can last only a few tens (or hundreds) of hours per year.

The capacity of the electrolyser must be capable to operate very quickly (within seconds) and should vary its power over a wide range (ideally between 0% and 100%), depending on fluctuations of excess electricity generation.

8.1.2 Power-to-gas-to-power

This concept (or power-to-power) covers the production of electricity from hydrogen or methane produced from methanation. This production can be achieved by a fuel cell, a combined heat and power (CHP) unit or a gas-fired power plant.

8.1.2.1 Fuel cell power

In the context of power generation from hydrogen or methane from power-to-gas, it is possible to use fuel cells or modified CHP units to use these gases.

There are still few manufacturers of high-power fuel cells greater than 100 kW (Table 8.2). In addition, molten carbonate fuel cell (Figure 8.1) or solid oxide fuel cells use generally hydrocarbons (natural gas or biogas).

Using methane from methanation decreases the overall yield.

8.1.2.2 CHP and conventional power stations

For CHP units, the maximum power is of the order of a few hundreds of kW (INNO Jenbacher with 531 kW_e and 630 kW_{th}, 2G Energie AG 150 kW_e and 172 kW_{th} or Caterpillar 1,250 kW) but the electrical efficiency is lower than for some fuel cells.

Gas-fired power plants are the only ones with high power of up to several hundred MW and a high efficiency (60% for combined cycle gas turbine [CCGT] units). Gas turbine suppliers have developed their equipment in order to fuel some models with hydrogen: GE Gas Power H-Class (50% hydrogen, 571 MW power, 64% efficiency) or Mitsubishi Power M701F (100% hydrogen by 2027, 440 MW_e).

The total yield, calculated from excess electricity generation, even if one of fuel cells or gas-fired power stations may exceed 50%, is still low compared to the direct use of hydrogen or methane produced. Whatever the technology, heat recovery and its use increase the overall efficiency.

Figure 8.1: A 2.8 MW fuel cell with two 1.4 MW modules (FuelCell Energy, Inc.).

Table 8.2: High-power fuel cells in 2022.

Supplier	Type	Power
FuelCell Energy, USA	MCFC	1.4 or 2.8 or 3.7 MW
Bloom Energy, USA	SOFC	300 kW
Ballard, Canada	PEMFC ClearGen II	1.5 MW
Nedstack, Netherlands	PEMFC PemGen	1 MW
Cummins, USA (former Hydrogenics)	Alkaline HySTAT	500 kW
Cummins, USA	PEMFC HYLYZER 200	1 MW

MCFC, molten carbonate fuel cell; PAFC, phosphoric acid fuel cell; SOFC, solid oxide fuel cell; PEMFC, proton exchange membrane fuel cell.

8.2 Power-to-gas economic evaluations

Given the still exploratory nature of existing facilities, it is difficult to give a cost that would be representative of mass production of electrolysers or fuel cells. However, the cost of existing equipment gives an estimate of the financing needs of existing or planned projects in the short term or medium term.

However, the COVID pandemic and war in Ukraine in 2022 have disrupted the market (extraction of raw material, reduced production, high transportation costs etc.) leading to a reduced supply and/or a drastic price increase.

8.2.1 Cost of operational facilities

Some basic components such as compressors or fuel cells are already produced industrially, sometimes in small batches. On the other hand, high-power electrolysers or methanation units remain as prototypes or pre-series, and the cost of which, often part of an evaluation programme, cannot yet be really estimated for future products.

8.2.1.1 Cost of an installation

An estimate made in 2022 for a 1 MW electrolyser by the Oxford Institute for Energy Studies [1] summarising different assumptions is indicated in Table 8.3.

Table 8.3: Split costs of a power-to-gas electrolyser in %.

Technology	Alkaline	PEM	SOEC	AEM
Stack	20	15	34	48
Electronics	15	10	6	15
Gas conditioning	15	15	30	18
Balance of plant	50	60	30	19
Cost in US $/kW (2019)	*540–900*	*667–1450*	*2300–6667*	*>931*

In 2021, The Fraunhofer Institute for Solar Energy Systems ISE evaluated the cost of low-temperature electrolysers in 2030 [2]:
- Alkaline electrolyser stack: €86/kW for 10 MW power (cathode = 24%, anode = 20%, separator = 12%)
- PEM electrolyser stack: €217/MW for 5 MW power (cathode = 14%, anode = 24%, MEA = 32%)

With those assumptions, the total cost of an electrolyser and balance of plant will be:
- Alkaline electrolyser: €726/kW for a 10 MW unit and €444/kw for a 100 MW unit
- PEM electrolyser: €980/kW for a 5 MW unit and €500/kw for a 100 MW unit

8.2.1.2 Cost of pilot projects

Although the costs of projects carried out or in progress include investments that are unlikely to be necessary for an industrial installation, they nevertheless allow to give a range of investment (Table 8.4).

Table 8.4: Overall costs of 2022 planned units.

Country	Project	Electrolyser	Power (MW)	Cost
France	H2VN Normandy	Alkaline/Hydrogen pro	2×100	€230 million
Norway	Yara	PEM/Sunfire	24	US $28.2 million
Finland	P2Xsolutions	Alkaline/Sunfire	20	€50 million
Germany	Lubmin	N.A.	50	€200 million
Germany	Oberhausen	PEM/Siemens Energy	20	€30 million
Denmark	Skive	Alkaline/Green Hydrogen	24	€30 million

This comparison shows the lack of relation between installed electrolyser power and global costs.

In the costs of these programmes (data from different actors), all installed elements are taken into account, sometimes even external infrastructure, wind turbines, fuel cells, CHP units etc.

8.3 Business model for power-to-gas

The profitability of a power-to-gas installation depends on many variables, such as economic (electricity and equipment prices), technical (technologies used), strategic (which option to implement), legislation and subsidies. All these factors will influence the overall costs of this technology.

8.3.1 Analysed system

The economic analysis of a power-to-gas unit can be based on a three-part division (input, process and output) whose different components vary according to the technology and the objective: hydrogen, methane or electricity production (Figure 8.2).
The cost of a specific system should consider the following:
- price of electrolyser, depending on the technology (alkaline, PEM or solid oxide electrolyte cell [SOEC]);
- price of compressor(s) if needed;
- price if hydrogen storage unit if needed;
- price of methanation unit production of methane is considered;
- price of injection unit (hydrogen or methane) if needed;
- price of transportation (trailers) if needed; and
- price of infrastructure (buildings, piping, cabling, electronics etc.).

The return on investment should also be related to the consumable (electricity, water, CO_2, complementary heat) and operational costs (OPEX) and to the market price of hydrogen or methane for the users.

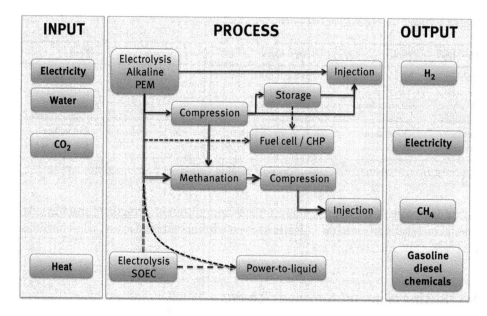

Figure 8.2: Potential systems for the power-to-gas approach.

8.3.2 Technical and economic analysis

The economic viability of a power-to-gas unit should be analysed considering the capital (CAPEX – CAPital EXpenditure) and operational (OPEX – OPerational EXpenditure) expenditures. As of 2022, the cost calculations for existing or future installations or projects vary from one unit or from one source to another depending on the assumptions for the costs of input, process and output considered (see Tables 8.3 and 8.4).

8.3.2.1 Levellised Cost of Electricity generation

Levellised cost of electricity (LCOE) represents the cost of generating electricity according to the technology used, allowing a cost comparison. The essential elements are the costs of the equipment, operation and maintenance as well as the annual operating time. Fuel costs (for conventional power plants: natural gas, fuel oil or coal) are zero for solar and wind.

The war in Ukraine in 2022 with the reduced delivery of Russian gas has disrupted the electricity and gas markets although the trend was already seen at the end of 2021: the average electricity price in Germany on EPEX SPOT between October 2021 and September 2022 was €200/MWh with peak up to €600/MWh. As a comparison between January 2019 and January 2021, the average price was about €40/MWh.

Thus, **LCOE** data available in 2022 reflect the 2021 situation (Table 8.5).

Table 8.5: LCOE of electricity (€$_{cent}$/kWh).

Technology	2021	2030	2040
CCGT	8–13	9–18	9–25
Nuclear (data: IEA)	6.5–15	7–12	6–12
Solar PV-utility scale (>100 kW)	5–10	4–7.5	2.5–6
Wind onshore	4–8	4–7.5	4–7
Wind offshore	7–12	6.5–11	6–10

Source: Fraunhofer Institute for Solar Energy Systems ISE; GWEC, Global Wind Energy Council; Lazard's LCOE Analysis-Version 15.0.
Note: Data from Fraunhofer Institute for Solar Energy Systems ISE, levellised cost of electricity, June 2021.

The 2021 data show the LCOE of the existing equipment. Projection for 2030, 2040 or 2050 will be influenced by the future cost of wind turbines, solar modules and infrastructure (cables, transformers etc.) which could increase drastically. Other factors will influence the LCOE of CCGT (gas and CO_2 pricing), being fairly stable up to 2021 or nuclear (maintenance for old plants) with continuously increasing LCOE.

Other factors like subsidies, local cost of equipment and manpower have an influence on LCOE which, for a detailed study, should be calculated for each country.

8.3.2.2 Grid parity

The comparison of the grid electricity price (depending on the country, mainly nuclear, coal or hydraulic) with that from renewable sources without subsidies gives an indication of the cost evolution. This was due to the decreasing price of solar modules and more powerful wind turbines associated with large-scale manufacturing units. Depending on national policy and electricity mix, grid parity should be analysed for each country. In China, for example, although coal contributes to about 70% of electricity production, solar PV (photovoltaics, representing only 4%) has reached grid parity in some areas in 2021 (US \$49.3/MWh) and it will be nationwide in 2023. In Germany, where electricity price is one of the highest in Europe (€$_{cents}$ 30/kWh in 2020), grid parity was reached in 2018 even for household PV and batteries.

8.3.2.3 Electricity price for power-to-gas technology

As the electricity used by the power-to-gas units should theoretically be the one in excess, the cost per kWh varies in a very wide window and could be even negative in some cases. Power-to-gas units can also negotiate guaranteed low prices insofar as they offer an alternative to the disconnection of PV or wind farms in the event of overproduction. The trend to consider hydrogen as a commodity, i.e. large-scale production for industry use and not using excess electricity, means taking a risk to pay high kWh prices, especially since beginning of 2022 where those prices soared, making eventually hydrogen production costs too high compared to pre-2022 projections.

Cost of other inputs

The cost of consumables includes water and carbon dioxide. The cost of water can be considered as negligible compared with that of electricity, and the same may be applied to today's carbon dioxide price (however, depending on the carbon market price evolution) and the heat needed by the SOEC electrolysers. However, water for electrolysis may be an issue in some countries where the availability is low or when it is competing with other needs like agriculture, for example.

8.3.3 CAPEX

Depending on the followed path, the investment of power-to-gas plant consists mainly of electrolysis, methanation, hydrogen storage, compression and injection into the natural gas network. In addition, the costs of real estate and investments and other auxiliary devices such as pipelines and gas conditioners contribute to the total CAPEX. For hydrogen injection in the natural gas network, the CAPEX is determined by the costs of the different components needed:

$$\text{CAPEX} = \Sigma \, \text{costs} \, (\text{electrolyser} + \text{injection unit} + \text{infrastructure} + [\text{storage}/\text{compression}])$$

The infrastructure covers the costs of building, piping, electronics etc. and storage or compression, depending on the characteristics of the electrolyser (output pressure) and the production capacity versus injection rate for an eventual storage.

The main cost contributor is the **electrolyser** with a range of, depending on the power, efficiency, production rate and the technology, PEM being still more expensive than alkaline. In 2021, as a guideline [3], CAPEX for an alkaline electrolyser was estimated to be €1,000/kW in 2020, €750/kW in 2030 and €400/kW in 2050 (values are respectively 1,000/500/250 for PEM electrolyser and 2,000/1,200/700 for SOEC).

The electrolyser operates continuously as long as excess renewable electricity is available, and a buffer **storage** is sometimes necessary. Storage pressure depends on the output pressure from the electrolyser and/or on the next step (injection into the natural gas network requires a lower final pressure than storage in a trailer or for transportation) and has an influence on the compressor cost.

Very few data are available for costs of **methanation** unit as the existing ones are demonstrators. However, it seems that biological methanation should be less expensive than thermochemical. The IEA estimates the costs to about US $850/kW depending on the size of the unit.

The 2021 International Energy Agency (IEA) report *Net Zero by 2050* estimates the CAPEX using low-temperature electrolysers to about US $835–1,300 /kW$_e$ in 2020, US $255–515/kW$_e$ in 2030 and US $200–390 kW$_e$ in 2050.

8.3.4 OPEX

The main cost of producing hydrogen is one of electricity. Two different ways of electricity purchase can be considered: on the spot market and through a long-term contract from a renewable energy producer (wind or solar).

The other influencing factors are:

- electricity consumption,
- consumables (water, carbon dioxide etc.),
- labour cost,
- depreciation and replacement and
- other OPEX related to auxiliary equipment (storage).

All of them depend on the capacity of the power-to-gas unit, the annual running time, the amortisation scheme and duration.

Fixed OPEX refers to direct labour, administration, insurance, taxes and maintenance. For the electrolyser, **variable OPEX** refers, for example, to annualised stack replacement (function of the lifetime) which can represent up to 60% (PEM), 50% (alkaline) and 60% (SOEC) of the plant.

Usually, the OPEX is estimated for a specified power of electrolyser and amortisation time. The Fuel Cell and Hydrogen Joint Undertaking has calculated it as about 10% for a 10 MW system (alkaline electrolyser) and 20 years lifetime [4]. A more recent study has been done by Deloite [5].

CAPEX and OPEX of electrolysers should be balanced in order to optimise efficiency, durability and cost: for example, an efficient stack reduces OPEX (less electricity needed per kg hydrogen)

8.3.5 Revenues

The operation of a power-to-gas unit derives its revenues from the sale of hydrogen or methane produced and eventually electricity (power-to-gas-to-power (P2G2P)). The oxygen and the heat produced can also be used for other processes.

The injection of hydrogen or methane gives advantages to the electricity grid (less "losses" of renewable electricity, stabilisation of networks etc.). For this reason, at least for a first phase, methane injection should benefit from a special tariff as the production costs (US \$0.1–0.5 kWh depending on the assumptions) are expected to remain higher than the natural gas costs (US \$0.02–0.03 kWh).

For industry, the price of methane or hydrogen produced does not have any real added value other than their purity which may be an important factor in some fields. In this case, hydrogen or methane produced should not be injected into the grid but follow a separate marketing circuit.

8.3.6 Levellised cost of hydrogen

The levellised cost of green hydrogen generation (Table 8.6), directly produced by onsite electrolyser) is dependent on many factors:
- electrolyser technology (alkaline, PEM, AEM or SOEC),
- electrolyser power,
- output pressure (without or with compressor),
- maintenance (stack lifetime),
- electricity cost and
- electrolyser runtime.

Table 8.6: LCOE of hydrogen (US $/kg).

Power	Electricity cost	Utilisation	Alkaline	PEM
20 MW	US $30/MWh	75%	2.64	3.20
20 MW	US $60/MWh	75%	4.86	5.75
100 MW	US $30/MWh	75%	2.53	3.05
100 MW	US $60/MWh	75%	4.75	5.59

Source: see *Lazard's LCOE of hydrogen Analysis-Version 20* for complete data set.

The IRENA 2022 report *Global Hydrogen Trade to Meet the 1.5 °C Climate Goal* gives the LCOE of hydrogen depending on the country (green electricity generation technology used: solar or wind and load of electrolyser): for Chile, an average of US $0.5–1.0/kg, about US $1.0–1.5/kg for Germany and US $1.5–2.5/kg for the USA.

The price of hydrogen is dependent on the production technology, and the volumes produced, the lowest being either from coal or reforming natural gas through large-scale plants. The 2021 IEA report *Net Zero by 2050* estimates the global hydrogen needs to about 212 million tonnes in 2030 and 528 million tonnes in 2050.

Hydrogen price is dependent on the technology used (Table 8.7 and Figure 8.3) to produce it. The variable is the price of the "raw" material like natural gas, coal or electricity for electrolysis. In the fourth quarter 2021, in the Netherlands, production costs of hydrogen were €507/MWh with alkaline electrolyser, €600/MWh with PEM electrolyser and €221/MWh with steam methane forming, all these are valid till December 2021, and the costs of electricity and natural gas have a great influence (in September 2021, costs were respectively €248/MWh, €300/MWh and €133/MWh).

8.3.7 Abatement of hydrogen

The carbon abatement of hydrogen is the reduction/decrease in the potential of CO_2 emissions by the production of green hydrogen. This is especially valid for hard-toabate industries generating actually large quantities of CO_2: steel, cement, chemicals, feedstock, for example. The abatement impact of hydrogen (emission reduction for each kg of hydrogen used) is determined by a combination of the CO_2 footprint when generated and the emissions when used.

The abatement factor can be calculated not only for the industry but also for transportation, for example. In the steel production, the blast furnace produces about an average of 1,600 kg of CO_2 for each tonne of steel, whereas the direct reduction of iron process emits 25 kg of CO and 25 kg of CO_2 while using 50 kg of hydrogen leading to an abatement of 31.5 kg CO_2/kg H_2 ([1,600–25]/50).

Table 8.7: 2021 hydrogen price comparison (US $/kg).

Technology	2021	2021 projection for 2030	2021 projection for 2050
Grey hydrogen (natural gas)	2–6	0.6–2.4	1.0–3.4
Grey hydrogen + CCS	6–10	1.0–3.4	0.9–2.1
Brown hydrogen (coal)	2–3	1.5–4.1	2.4–6.5
Green hydrogen (electrolysis)	10–13	1.2–3.9	1.0–3.4

Source: International Energy Agency (IEA).
Note: However, all those long-term estimations are based on 2020/2021 assumptions. Already in July 2022, the price of green hydrogen reached US $16.8/kg. Natural gas price increase will also affect the cost of grey hydrogen. All those variabilities leave an uncertainty about the different scenarios for 2030 or 2050.

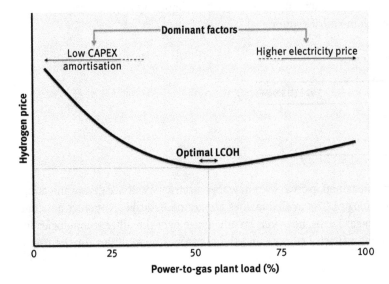

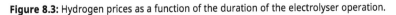

Figure 8.3: Hydrogen prices as a function of the duration of the electrolyser operation.

In 2021, the *Hydrogen Council* estimated the global hydrogen demand to be 140 million tonnes in 2030 (at least 50% green) and 660 million tonnes in 2050 (up to 80% green hydrogen). If this trend is confirmed from 2021 to 2050, 80 Gt of cumulated CO_2 emissions will be abated.

The associated abatement cost can be calculated as the cost reduction of each tonne of abated CO_2. Besides hydrogen produced from renewable electricity, natural gas reforming and CCS (carbon capture and storage) contribute to CO_2 emissions. The French governmental organisation *France Strategie* estimated, in 2022 [6], the abatement cost of renewable hydrogen to be €200/tonne CO_2 (assuming a very optimistic cost of green hydrogen to €3.5/kg) and €100 for reforming + CCS (2021 natural gas price).

8.3.8 Comparison with other electricity storage technologies

Different technologies are available for electricity storage. Their application will depend on the defined use, which could be grid stability, surplus storage, time shifting, resiliency (backup or reserve) etc. Pumped hydrostorage, compressed air energy storage, batteries (conventional where lithium technology is dominating, redox flow, NaS) and flywheels are the current options.

To compare the costs of storage (Table 8.8), the common metrics used are the investment costs and the LCOS (levellised cost of storage), covering the operations and maintenance during the lifetime of the installation [7].

Hydrogen storage depends on the volume to be stored, storage duration and storage technology (cavern, pressurised, liquified), where liquid hydrogen being the most expensive and cavern the most economical.

Table 8.8: LCOS in US $/MWh.

Technology	2021 (US DOE)
Lithium batteries	322–369
Hydrogen in cavities	98–241
Hydrogen pressurised	115–260

The increasing and sustained growth of renewable electricity will accelerate the adoption of storage technologies. The available ones are not necessarily competing but complementary. Power-to-gas keeps, however, an advantage over the other technologies for the (almost) unlimited capacity storage and the numerous possibilities for the use of hydrogen produced.

Power-to-gas financial approach and sustainability

The necessary investments to increase the renewable electricity ratio (up to 50%, 80% or even 100%) will include the storage of surplus electricity. The power-to-gas technology could absorb the expected large surpluses. The profitability is not reached yet but the costs (CAPEX and OPEX) will decrease due to the production of equipment in series and the increase of the overall yields. Before reaching this critical stage, an intermediate step through government support and a specific policy will be needed to put this technology on a path of profitability.

Currently, the main economic value of power-to-gas technology is the production of hydrogen or methane. However, the real benefits to the energy system are the development of local solutions (decentralised grids), avoiding costs of extension of electricity infrastructure (transmission over long distances), often promoted by governments or energy companies, and extending the use of the existing gas network.

8.4 Availability and prices of critical raw materials (CRM)

The energy transition is based among others on renewable electricity (wind and solar). Wind turbines, solar modules, electrolysers and batteries require components using critical raw materials (CRM).

The 2021 IEA report [8] lists metals like nickel, aluminium, chromium, cobalt, copper and rare earth elements (niobium, neodymium, dysprosium etc.). The mineral consumption of energy sector will increase drastically mostly accelerated by the demand of wind turbines, solar modules, electricity networks, electric vehicles and battery storage: for copper from about 22% of demand in 2020 to an expected 30–45% in 2040; for nickel, respectively, from 8% to 30–60%; and for rare earths from 15% to 42% (IEA 2021 data, the second figure for 2040 to be in agreement with the Paris COP objectives). The high demand and the COVID and war in Ukraine disruptions have led to an increase in prices (Figure 8.4),

8.4.1 Dependence on a few countries for extraction

Some CRMs are extracted only in a few countries or controlled by a few companies leading to a strong dependence (Table 8.9). Some of those countries have also unstable political or social conditions.

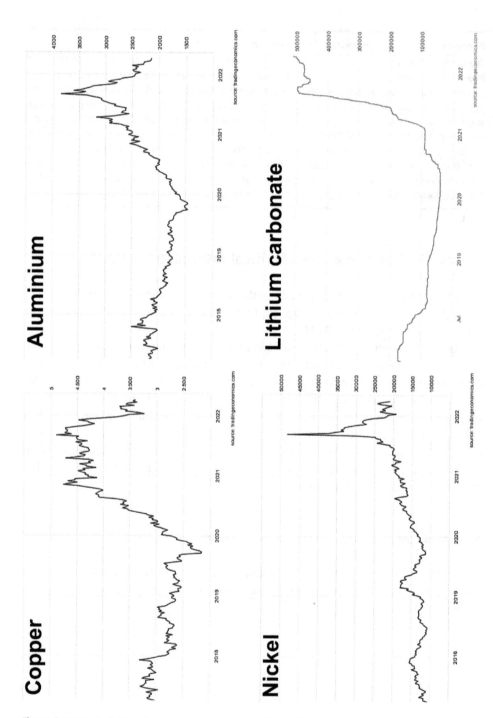

Figure 8.4: Price evolution of some energy sector materials (trading economics).

Table 8.9: Main suppliers' contribution to some CRM (2020 data).

Country/material	Nickel	Copper	Cobalt	Rare earths (global)	Iridium	Platinum	Ruthenium
China		12%		60%			
Chile		28%					
Russia	11%		4%	46%			
Indonesia	33%						
Democratic Republic of Congo			69%				
South Africa					85%	70%	93%

Chinese companies, for example, control 80% of rare earths worldwide production through mine acquisitions.

8.4.2 Dependence on China for processing

Processing of materials is energy intensive and has an important greenhouse gas intensity. For rare earths, China controls about 90% of worldwide refining capacity in 2022 (Table 8.10). China has a market share of over 50% for numerous materials: cobalt refining represents 65% of world's demand, tungsten represents 85% etc. Western countries imported critical materials without real ecological or environmental criteria.

Table 8.10: China contribution to materials treatment and second country (2020 data).

Material	Nickel	Copper	Cobalt	Rare earths (global)
China	35%	40%	65%	87%
Indonesia	15%			
Chile		10%		
Finland			10%	
Malaysia				12%

In the frame of Chinese policy to increase the internal market, already in 2010, the exportation of some materials was restricted.

8.4.3 Availability versus needs

The race for CRMs is running in all developing countries to cover the needs of their industries.

The EU has defined a list of CRM that is constantly updated. It shows the great dependence on supply (Europe has given away mining long ago) and treatment (subcontracted to countries with less severe environmental legislation and lower labour costs). In 27 countries of EU, mining represents only 5% of the needs. In 2020, the USA imported 100% of 14 minerals from the critical list and depended on 75% for 10 other minerals.

8.4.3.1 Is large-scale mining in Europe or in the USA a viable solution for a reliable supply?

The long administrative requirements, financial and environmental aspects, and equipment availability can lead to a duration of up to 20 years before being operational. The other issue is the public acceptance and social impacts, especially in the western countries due to the bad reputation of the sector (characterised as "dirty, dusty, dangerous"). Displacements of households and/or land acquisitions are not always accepted: in 2021, the Rio Tinto project for a lithium mine in Serbia has been followed by strong protests from the population, and the permit was cancelled by the Serbian government. In Australia, another lithium project is threatening a sacred spring used by the Hualapai people. Even if new mines could open, it will take years to cover the initial costs and they will compete with the existing one already amortised. Another unknown issue is the availability of enough material from these potential new mines compared to the investments and the needs: the ore concentration should be economically viable and the quantities to mine cover decades of production which is rarely the case. Often the projects are based on very optimistic production potential figures and do not take into account the market requirements compared to the expected production. The last step (treatment) has also to be conducted in the extracting country, i.e. more investments and starting the learning curve. However, considering the projects announced or planned compared to the existing facilities (mining or processing), very little change is to be expected in the near-term or mid-term.

The **recycling** of batteries, PV modules and decommissioned wind turbines, if done locally, will contribute to the reduction of importation and dependence. Recycling depends on the level of collection, market price and local processing units.

8.5 CAPEX and OPEX in a disrupted world economy

Most prospective CAPEX/OPEX studies, estimations or projections have been issued in 2021, before the war in Ukraine. Added to this, the world economy has not recovered from the COVID pandemic (Chinese economy is still on the low side), and transport costs have also soared. High inflation (up to or over 10% for some countries in 2022), energy and critical material prices have changed the game.

With those parameters, defining CAPEX/OPEX for future projects or for existing installations is less reliable, and based on 2022 data could impact the financial viability of existing or future projects. In 2021, some studies have set green hydrogen price at US $1.5/kg in 2025/2030 (in July 2022, it spiked at US $16.8/kg)!

References

[1] Oxford Institute for Energy, Cost-competitive green hydrogen: how to lower the cost of electrolysers? January 2022.
[2] Fraunhofer Institute for Solar Energy Systems ISE, Cost Forecast for Low Temperature Electrolysis, October 2021.
[3] IRENA, Green Hydrogen Cost Reduction, 2020.
[4] Fuel Cell and Hydrogen Joint Undertaking (FCH–JU), Commercialization of Energy Storage in Europe, Final Report, 2015.
[5] Deloitte, Fuelling the Future of Mobility, January 2021.
[6] France Stratégie, Les coûts d'abattement, Partie 4-Hydrogène, Mai 2022.
[7] NREL, Storage Future Studies, May 2021.
[8] IEA, The Role of Critical Minerals in Clean Energy Transition, March 2022.

9 Role of power-to-gas in energy transition

With the acceleration of the global warming, the 2022 energy crisis and the climatic catastrophes (droughts, flooding, fires etc.), the necessity to move to clean energy sources is more than critical. The associated needed measures are for example reduction of energy consumption and change of nutrition (less meat for example). If 2020 saw a decrease of greenhouse gas emissions (CO_2, methane, nitrous oxides, chlorofluorocarbons etc.), due mainly to COVID lockdowns, emissions started to increase in 2021, accelerated by the high gas prices leading to more coal use for electricity generation. To limit the global warming to an acceptable level at the end of the century, the Paris agreement, adopted in 2015, set the level to ideally well below 2 °C above pre-industrial levels. However, in mid-2022, the World Meteorological Organization estimated [1] that the probabilities are 50/50 so that an increase of 1.5 °C could be reached by 2026. Based on actual trends, by 2100, temperature increase could be in the range 3 to 5 °C, depending on the simulation.

The decarbonation should reach all economic sectors (industry, transports, services, households, buildings etc.). The development of renewable energies has to be increased and speeded up associated with energy savings in those sectors. For electricity, wind and solar will be predominant (it could reach 90% in 2040 in Europe). The move to an (almost) all electric society could not be necessarily reached but the increase of renewable electricity will lead to surpluses to manage. The power-to-gas approach can be part of the storage of those surpluses.

9.1 Impact of power-to-gas on energy systems

The power-to-gas concept goes beyond the national framework. For a multistate solution, harmonisation of the various national regulations is necessary in order to facilitate the approval of installations as well as the production, storage, import/export or use of hydrogen and injection into the gas network naturally.

9.1.1 Legislation and regulations

9.1.1.1 Energy policy targets
The implementation of power-to-hydrogen can help address a variety of climate and energy-related objectives:
- reduction of greenhouse gas emissions,
- increase of the share of renewable energies and
- reduce energy consumption by moving to more energy-efficient solutions like electric vehicles, heat pumps and electrification of industrial processes.

https://doi.org/10.1515/9783110781892-010

The different strategies defined by many countries for an energy transition are often part of a series of regulations concerning energy, transportation, housing, industry or agriculture.

The development of the power-to-gas technology will influence many sectors of the economy and should be integrated in the regulatory and legislative framework.

Legislations should follow the same pattern by removing barriers hindering the use or extension of power-to-gas like increasing gas storage capacity or recognising indirect electricity storage as gas, specifying a percentage of hydrogen in the natural gas network based on ongoing experimentations. In the implementation of results of projects like CertifHy (Definition of Green Hydrogen), HyReady (Preparing Natural Gas Networks for Hydrogen Injection) or the standardisation of systems and devices for the production, storage, transport, measurement and use of hydrogen through ISO, CENELEC should also facilitate the penetration of power-to-gas technology.

In 2021, the European Union adopted a *Hydrogen and Decarbonised Gas Package* including in the proposal future hydrogen networks, interconnections with third countries, storage and terminal. In August 2022, the USA passed the *Inflation Reduction Act* that included hydrogen Production Tax Credit.

The European *Electricity Directive 2019/944* considers power-to-gas as "energy storage" in the electricity system including integration of the gas and electricity sectors. The project STORE&GO (2016–2020) involved three power-to-gas concepts and demonstrators in Germany (Falkenhagen), Switzerland (Solothurn), and Italy (Troia). The objectives were to show the synergies between electricity and gas for transportation, storage, and use. Following the Fuel Cells and Hydrogen Joint Undertaking (FCH-JU) started in 2008, in 2022, the EU's *Clean Hydrogen Partnership* (public–private investments) with a budget of €300 million covered 41 research projects in the production, transportation and storage of hydrogen.

9.1.2 A new architecture for energy networks

9.1.2.1 A separation between production, transmission and distribution of electricity

A decentralised approach should encourage local experiments carried out by other entities (municipalities, communities, alternative energy providers etc.). A separation of production, transmission and distribution can only be favourable to a competitive market with all producers considered as equal for access to the electricity or gas grid.

In many countries, several electricity or gas producers and operators of transmission/distribution networks or local actors are active, while in others like in France, for example, a situation of a de facto monopoly exists (the energy producer EDF manages the transmission and distribution through its subsidiaries RTE and ENEDIS) causing a distortion of the market.

9.1.3 Need for decentralisation

The centralised approach that still characterises the strategies of many producers and suppliers of energy (electricity or gas) runs against the evolution of electricity generation, which is increasingly decentralised: wind farms or photovoltaic installations scattered throughout the territory, combined heat and power (CHP) units of all sizes etc.

Distributed generation can be characterised by:
– local production or storage of electricity,
– located near or at load centres,
– grid connected or isolated and
– automated grid management

The power-to-gas technology can be integrated in such a scheme through local production of hydrogen or methane as well as a local use or if necessary, injection of the surplus not used into the natural gas network.

9.1.3.1 Microgrid

Local electricity management should allow a better optimisation of production/consumption flows. The gains expected from this approach include short-distance transmission (less losses), reduced transmission or distribution line loads, local network stability easier to manage than large networks and economic value by avoiding the construction of new transmission lines.

Depending on the size of these local networks, as microgrid covering a street or as minigrid for larger areas or small agglomeration, local production and consumption must allow for appropriation by users. A local power-to-gas unit, possibly coupled with a biogas unit, would first feed the nearby users (hydrogen, methane, electricity and heat), and the surplus could either be stored or injected into the electricity or natural gas grid. The power-to-heat approach could complement the microgrid. A microgrid also requires a control and communication infrastructure in order to dispatch and store locally the energies produced and only after injecting them into the respective networks.

Experimentation of a microgrid with storage

The "Strombank" (power bank) project in Mannheim, Germany, conducted between 2014 and 2016 tested a microgrid based on photovoltaic electricity generation or micro-CHP and involved 18 participants (14 dwellings and 4 tertiary activities). Storage was provided by batteries (100 kW/100 kWh), and each user had an "account" where he/she could store his/her surplus electricity and use it later. If the account was empty, it was possible to "purchase" virtually electricity from other participants. With this approach, self-consumption increased from 30% to 60–80%.

Many other microgrid projects are either planned or under development.

In India, a project awarded in 2021 to National Thermal Power Corporation Limited (NTPC) in Andhra Pradesh involves a 250 kW SOEC electrolyser from Bloom Energy to be fed by a floating solar project. Hydrogen will be stored under high pressure and, when needed, converted into electricity for a guest house by a 5 kW fuel cell. In 2022, the New York State Energy Research and Development Authority

(NYSERDA) developed a program to support locally owned microgrids that use green hydrogen fuel cells. In Australia, in the northern Queensland, a microgrid project planned for 2024 will use electricity from a 8 MW solar farm, 20 MWh of battery storage and a 1 MW hydrogen plant for residents. In Europe, the program ALPGRIDS (France, Switzerland, Germany, Italy and Slovenia) will implement seven pilot projects. The Italian site in Savona is planning an hydrogen CHP unit. The European REMOTE program (2018–2023) will demonstrate the feasibility of hydrogen and fuel cells for remote areas. Projects are taking place in Greece, Norway and Spain. Renewable electricity will be used to produce hydrogen and, when needed, converted into electricity.

The integration of hydrogen in microgrids (from renewable electricity) allows more flexibility for storage, eventually associated with batteries, and shows more flexibility (electricity with a fuel cell, household equipment using hydrogen etc.) and power reliability.

9.1.3.2 Virtual power plant

This concept, based on the convergence of micro- or minigrids, covers the management of a set of power and heat generation units forming a network and spreading over a territory (district, agglomeration) in order to optimise production and local consumption, the whole being managed as a single central (virtual) unit (Figure 9.1).

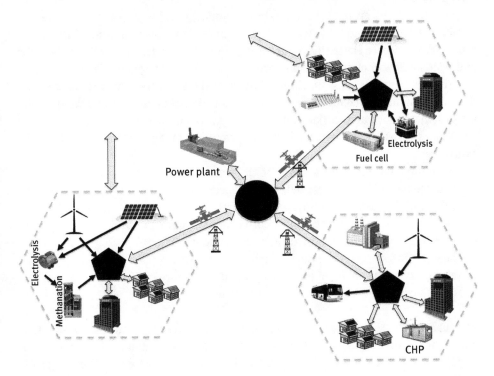

Figure 9.1: Local virtual power plant.

The management of different units (start-up, power to be delivered etc.) allows an optimal use of available resources, whatever the type of energy available: renewable (wind, photovoltaic, small hydro) or non-renewable (natural gas). Local management can also be extended over time with local storage controlled by production and consumption forecasts.

The overall optimised domestic approach to energy should be based on the smallest unit, a **smart home**, where some appliances can communicate bidirectionally with the **smart meter** and others are optimised for low power consumption (e.g. lighting with the presence of detector, shutters or blinds controlled by brightness and outside temperature). This management can only be done if the price of electricity to the consumer is updated at a frequency sufficient to reflect the price on the electricity exchange market. In this way, a low price of kWh would lead to the start of domestic appliances (**smart appliances**) having a high consumption (washing machine, dishwasher etc.), which can be controlled by the smart meter. Consumer awareness would be possible if the consumer can check easily this information on display screen or smartphone, for example.

9.1.3.3 Smart grid

It refers to the integration of information technology (IT) (sensors, communication, monitoring, control) into the power grid. It is supposed to deliver efficiently sustainable, economic and secure electricity. All users and equipment connected to this grid and the bidirectional flow of electricity and information could be managed in real time.

The smart grid is only "intelligent" through the software structure for meeting the user demand, managing energy from any source including solar and wind, avoiding system overload and working autonomously when needed.

From the last years' experiments, it appears that many challenges remain to be overcome:

- definition of a common standard for all involved equipment,
- management of a very large data flow collected,
- real-time electricity prices (dynamic pricing),
- availability of appliances with standard communication protocol,
- vulnerability of the infrastructure (hacking, manipulation),
- heavy IT infrastructure requiring high-level expertise,
- electricity consumption of the smart grid infrastructure itself and
- overall cost of deployment of smart grid structure.

Although smart grid evaluations have been conducted in some countries (France, Germany, the UK, Spain, Japan, the USA etc.) and are the subject of many laboratory studies [2], it is not clear whether the benefits will cover the installation and running costs of the current projects.

From 2007 to 2021, the EU has supported 407 projects for a total investments of €3 billion.

However, the principle of the smart grid is to provide a decentralised, autonomous and intelligent energy management system but it is still often planned or conducted by the large energy suppliers in a top-down and still locally centralised approach.

Data protection and network security

The technologies described allow operators to collect very detailed consumer data (high-frequency data transmission) which would allow to know the habits of each user. Data mining could, for example, allow this data to be used for targeted marketing strategies.

A regulation of the rights of users to manage the use of generated data must be part of the data protection approach. The European Commission has launched a regulatory approach.

The connectivity envisaged also opens the door to hacker's attacks of these networks. The numerous examples of pirated equipment, companies or organisations show that the absolute protection is non-existent.

9.1.3.4 Blockchain technology and power-to-gas

The blockchain, often associated with the Bitcoin, a virtual cryptocurrency invented in 2008, is an approach that can be applied in many sectors including energy.

The basis of blockchain technology is a direct digital transaction ("smart contracts") between two actors where the data, gathered in a block, are secured (validation, transmission and storage) using a peer-to-peer network involving all users of the network. This transaction is finally integrated with others in a chain (blockchain) and retained by all users. Compared to traditional transactions, the blockchain makes it possible to avoid any intermediary, especially if a cryptocurrency (Bitcoin, ether) is used (Figure 9.2).

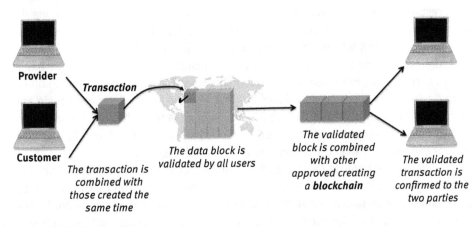

Figure 9.2: Blockchain principle.

Applied to electricity, for example, salesperson (producer) and user each having an account on the same platform could agree on a volume of exchange and on price.

This transaction, which can be immediate or linked to certain criteria, then passes into the blockchain process to be validated.

Current experiments in the field of energy [3] cover only certain aspects of the blockchain as there is always still one or more intermediaries involved:

In South Africa, the Bankymoon project, based on a prepaid smart meter, uses the Bitcoin. Each counter is linked to a digital account credited to each payment of the user who pays according to his/her means. The suppliers are thus assured of payment, weak point in South Africa. Bankymoon also targets schools that require reliable electricity supply through a system allowing any person or company in the world to perform a donation in Bitcoin. Founded in 2014, it is still active in 2022.

In 2016 in Germany, the BlockCharge project from the RWE Innovation Hub simplified the payment of electric vehicle loads by developing a direct smart contract between the electricity supplier and the user via a special Smartplug for all terminals.

For the port of Rotterdam, Distroenergy has developed in 2020 a platform to buy and sell the port's roof photovoltaic electricity. High-frequency trading and blockchain allowed to increase local consumption, optimise returns and reduce emissions.

The Spanish energy supplier Iberdrola started in 2019 a pilot project to inform corporate customers and other users about delivered renewable electricity (guaranteed origin), and to trace the source by linking data from the plants to consumers. Another Spanish energy supplier, Acciona Energía, developed a digital platform *GreenChain* to track and document renewable energy production. In 2021, the project *GreenH2chain* used the same approach but for green hydrogen production.

Peer-to-peer digital transactions can lead to drift in terms of smart contracts (consumer rights, litigation management etc.) or in their implementation, hence the need for a regulatory authority. While all aspects of the transaction are carried out via the Internet, it remains that electricity, for example, needs a physical transportation and distribution network where other external actors intervene. Moreover, the financial part, if it is not done in a cryptocurrency (the Bitcoin has a very variable exchange rate vs euro or US$), must also involve a banking institution. We are therefore still far from a real peer-to-peer direct transaction that was the heart of the blockchain and that would be applicable to a microgrid. The main advantage that the major players in the world of energy currently see in the blockchain is simplified and, in principle, secures transactions and accounting (smart contracts) (Figure 9.3).

Blockchain technology can be used to control electrical networks via smart contracts. Depending on the conditions of the transactions, they automatically manage production, sale or storage ensuring a balance between supply and demand. In case of surplus production, the surplus is automatically stored by other entities (this storage can also be the subject of a smart contract) and if the demand exceeds production the stored electricity is used which can also be part of another smart contract.

Other than executing energy transactions, the blockchain can provide documentation of ownership, asset management, guarantees of origin of electricity or green certificates.

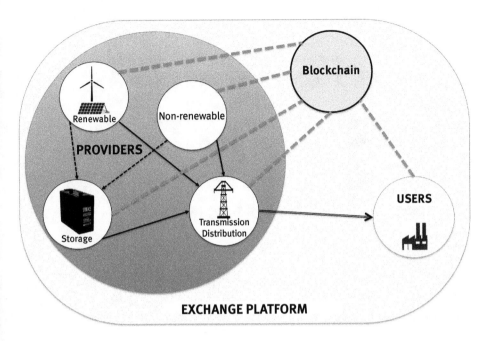

Figure 9.3: Blockchain principle applied to the electric grid.

9.1.4 Convergence of electricity and gas grids – sector coupling

If the power-to-gas main value is to "recover" surplus electricity, the resulting output (hydrogen or methane production) can be integrated into the natural gas network. So far, the two grids were using two different sources and were separated. The coupling of these two networks (electricity and natural gas) is a key technology for an energy network characterised by a high volume of intermittent energy sources (Figure 9.4).

This synergy will lead to a better optimised management of these two networks in order to maximise the gains of each: production of hydrogen or methane from the electricity grid and use and/or storage of these gases mixed with natural gas. This would allow the later use of these gases for households, services, transport, industry or electricity production (CHP, fuel cell and gas power plant).

9.2 Power-to-gas technology: a contribution to the protection of the environment

The impact of climate change, resulting in meteorological disruptions, which makes it difficult to assess the extent of change, increases the vulnerability of energy systems. Hydropower with more drought periods and the electricity network with

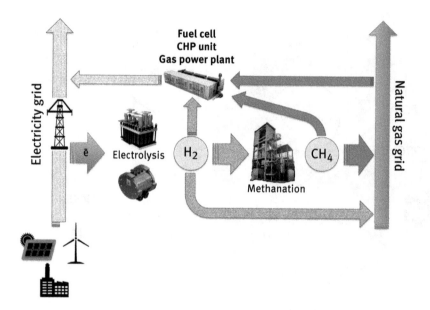

Figure 9.4: Convergence of electricity and gas networks.

more frequent hurricanes or storms will be strained. **Energy decentralisation**, whether for production, management or use, should make it possible to secure supplies by managing resources locally.

The local exploitation of the surplus electricity and its possible conversion to methane should make it possible to avoid the construction of new high voltage lines: this methane could be carried by the existing natural gas network for residential, tertiary, industry or agricultural sectors. It will also be possible to use it for gas-fired power plants, CHP units or fuel cells to produce electricity locally.

9.3 "Hydrogen civilisation" or electron civilisation?

The production of hydrogen in large quantities from renewable electricity and its use in almost all sectors has led to the concept of "hydrogen civilisation". The main thesis of this approach is the creation of a hydrogen infrastructure that would penetrate many sectors of the economy including transportation. The fascination of hydrogen, a simple element, periodically leads to a "renewal" of this idea.

Hydrogen and transportation

Considering the costs of fuel cell electric vehicle and of the necessary infrastructure (delivery, refuelling station), hydrogen is not well suited for transportation. All components of this approach are still expensive and energetically (well-to-wheel) not efficient. But for some captive applications like forklifts, for example, the economic advantages are more obvious than conventional solutions (batteries or natural gas-powered internal combustion engine).

The power-to-gas technology is a way to move out from non-renewable sources of energy. It does not mean going towards a "hydrogen society" but to open a new economic field based on hydrogen. The objective is not to build an energetic approach based upon and around one source of energy (in this case hydrogen), as the past has shown what relying only on one (coal first then oil or gas) can hinder development of other alternatives.

9.3.1 Will there be enough hydrogen?

For the current industry users, hydrogen demand will rise from about 95 million tonnes in 2021, with a few percent from renewable electricity, to 180 million tonnes in 2030 (IEA estimates with 62 million tonnes green hydrogen from electrolysis). Those additional 62 million tonnes of green hydrogen will require 248 TWh of renewable electricity (assuming 4 kWh/kg of hydrogen). As a comparison, in 2021, Germany's electricity generation from wind and solar was 162 and 919 TWh for the EU (wind, solar and hydropower). However, the global 2050 hydrogen demand for existing and new sectors like transportation is estimated to be 528 million tonnes (322 million tonnes of green hydrogen requiring 1,288 TWh)!

The share of renewable electricity in total generation is still fairly low (29% in 2021 and 61% in 2030, according to the IEA). To be characterised as green, hydrogen should be produced from renewable electricity. However, this electricity is first needed as electrons for direct use by the industry, households, services etc. In some large CO_2 emitting sectors (steel, cement, glass) or refineries, decarbonation can be reached by the use of hydrogen in the processes. In that case, hydrogen production for those applications can be justified. Other applications requiring large quantities of hydrogen should use only surplus electricity to produce it. Transportation, for example, can use directly electricity with storage in batteries (cars, trucks, trains, ferries etc.) although the same issue as the needed quantities of hydrogen arises: due to the important expected increase of electric vehicles, will there be enough processed lithium available?

The fall/winter 2022/2023 has shown how critical the supply of electricity can be influenced by weather conditions (less rain to supply hydropower), political situation (war in Ukraine, stop of Russian natural gas supply etc.) or technical issues (shut down of 50% of nuclear plants in France). This shows the necessity and importance to use only surplus electricity to produce green hydrogen.

The hydrogen illusion

The production of hydrogen is sustainable only if it is produced from (excess) renewable energies. Renewable electricity is already needed to reduce emissions from other sources (coal or natural gas). Directly using this electricity for transportation, for example, is the most efficient method even if the electric vehicle suffers from limitations such as relatively low effective range and long charging times: it can use up to 90% of the electricity produced, while the one using hydrogen for an on-board fuel cell will use only about 30% of the initial energy.

According to Ulf Bossel, an expert in this field, the "rush" towards an economy based solely on hydrogen [4] is not supported by reasons of energy efficiency nor by economic or ecological considerations. Hydrogen should be limited to sectors where it will be impossible to circumvent (e.g. the power-to-gas). Renewable electricity is better used in the form of electrons than hydrogen. The current energy structure linked to power-to-gas technology does not permit the direct use of hydrogen, which is still linked to too many uncertainties (concentrations to be used, equipment to be modified or replaced etc.). Methanation, on the other hand, opens up prospects for the exploitation of excess electricity without changes for users.

9.4 How would an efficient hydrogen strategy look like?

The power-to-gas approach can be easily integrated into the existing energy infrastructure. However, focus should be based not only on technology but on the role it can play in the energetic transition [5, 6]. Power-to-gas can be considered as a unique disruptive "storage" technology compared to other options (batteries, PHS (pumped hydrostorage), compressed air energy storage or flywheel) as it shows specific features:

- it is the only storage technology being able to "absorb" large volume of excess electricity versus batteries or PHS and
- it is the one whose "product" (hydrogen) can be easily stored over long time without losses and used in many economic areas.

Added to those advantages, power-to-gas can also contribute to reduce energetic dependence with mature basic technology (validated high-power electrolysers with up to 20 MW in 2022 and storage or injection into the natural gas network). The next step for its large-scale integration remains series production to reduce costs and increase capacity.

The current situation is characterised by too many projects (especially demonstrators), where often the wheel is rediscovered at a cost of time and money (the French project Jupiter 1000 has an objective to compare PEM and alkaline electrolysers. Those technologies have been already tested and validated in other countries). Interesting projects are also time limited, even if from the infrastructure and results point of view they could be operational much longer.

The development of power-to-gas to a large scale, the costs and technology optimisation would require:

- coordination of projects at the country level and for the EU at the EU level,
- centralisation of national, European or international R&D projects to avoid duplicates and allocate sufficient funding,
- speeding up the move to industrial scale and
- stopping the unnecessary competition between countries and within a country between the different regions.

The hydrogen (and power-to-gas), still seen as a contributor to the energy transition and the decarbonation of the economy, is however lacking serious investments to reach the step, where it will be an important part of the energy sector in terms of storage and use. Faced with huge energetic and environmental challenges, the progresses are still limited and the projects are not up to the real needs.

The power-to-gas concept, whether by hydrogen or methane produced, is an element of the energy transition and decentralisation of the energy system. Faced with energy and environmental challenges, the solution cannot be provided by a single technology, but power-to-gas can contribute to it.

References

[1] World Meteorological Organisation, Global Annual to Decade Climate Update, 2022.
[2] Vasiljevska, J. and Efthimiadis, T. Selection of Smart Grids Projects of Common Interest – Past Experiences and Future Perspectives, MDPI, March 2022.
[3] European Joint Research Centre, Blockchain Solutions for the Energy Transition, 2022.
[4] Bossel, U. The Hydrogen Illusion, Cogeneration and On-Site Power Production, March–April 2004, 55–59.
[5] Hydrogen Council, Policy Toolbox for Low Carbon and Renewable Hydrogen, November 2021 and Hydrogen for Net Zero, November 2021.
[6] IRENA, Geopolitics of the Energy Transformation, The Hydrogen Factor, 2022.

Acronyms

AC	Alternating current
CAES	Compressed air energy storage
CCGT	Combined cycle gas turbine
CCS	Carbon capture and storage
CEN	European Committee for Standardization
CENELEC	European Committee for Electrotechnical Standardization
CGH2	Compressed hydrogen
CHP	Combined heat and power
COP	Coefficient of performance (of heat pumps)
CSP	Concentrating solar power
CSPE	Contribution au Service Public de l'Electricité (France)
DC	Direct current
EEG	Erneuerbare-Energien-Gesetz (Germany)
EPEX	European Power Exchange
EPIA	European Photovoltaic Industry Association
ENTSO–E	European Network of Transmission System Operators for Electricity
ESO	European Standardization Organization
EV	Electric vehicle
FCEV	Fuel cell electric vehicle
GDL	Gas diffusion layer
GWEC	Global Wind Energy Council
GHG	Greenhouse gases
GNP	Gross national product
HVDC	High-voltage direct current
HHV	Higher heating value
ICE	Internal combustion engine
IEA	International Energy Agency
ISO	International Organization for Standardization
LCOE	Levelled cost of energy
LCOH	Levelled cost of hydrogen
LCOS	Levelled cost of storage
LH2	Liquid hydrogen
LHV	Lower heating value
LAES	Liquid air energy storage
LNG	Liquefied natural gas
LOHC	Liquid organic hydrogen carrier
MCFC	Molten carbonate fuel cell
MEA	Membrane electrode assembly
MH	Metal hydride
NGV	Natural gas for vehicle
O&M	Operations and maintenance
OECD	Organisation for Economic Cooperation and Development
P2G	Power-to-gas
P2G2P	Power-to-gas-to-power
P2H	Power-to-heat
P2L	Power-to-liquid
PEM	Proton exchange membrane

https://doi.org/10.1515/9783110781892-011

PEMFC	Proton exchange membrane fuel cell
PMG	Precious metal group
PHEV	Plug-in hybrid electric vehicle
PHS	Pumped hydrostorage
PSA	Pressure swing adsorption
PtG	Power-to-gas
PtH	Power-to-heat
PtL	Power-to-liquid
PV	Photovoltaic
RES	Renewable energy sources
SAIDI	System Average Interruption Duration Index
SAIFI	System Average Interruption Frequency Index
SMES	Superconducting magnetic energy storage
SMR	Steam methane reforming
SNG	Synthetic or substitute natural gas
SOEC	Solid oxide electrolyte cell
SOFC	Solid oxide fuel cell
T&D	Transmission and distribution
UCTE	Union for Coordination of Transmission of Electricity
YSZ	Yttrium-stabilised zirconia

Index

https://doi.org/10.1515/9783110781892-012

Printed in the USA
CPSIA information can be obtained
at www.ICGtesting.com
JSHW050015170524
63288JS00010B/161

9 783110 781809